Celestial and Stellar Dynamics

Although the field of celestial dynamics – the application of Newtonian dynamics to systems with a relatively small number of celestial bodies – is centuries old, it has been reinvigorated by the discovery of thousands of exoplanetary systems orbiting other stars. This textbook uses the properties of planetary systems, including our own solar system, to illustrate the rich variety of behavior permitted by Newton's law of gravity. The textbook then expands its view to examine stellar dynamics – the study of systems containing a very large number of stars or other celestial bodies. The different techniques used for celestial dynamics and stellar dynamics are compared and contrasted. However, throughout the text, emphasis is placed on the underlying physics that applies on scales as small as the Earth–Moon system and as large as a cluster of galaxies. It is ideal for a one-semester astrophysical dynamics course for upper-level undergraduates and starting graduate students.

BARBARA RYDEN received her PhD in astrophysical sciences from Princeton University. After postdocs at the Harvard-Smithsonian Center for Astrophysics and the Canadian Institute for Theoretical Astrophysics, she joined the astronomy faculty at The Ohio State University, where she is a full professor. She has 30 years of experience in teaching, at levels ranging from introductory undergraduate courses to advanced graduate seminars. She won the Chambliss Astronomical Writing Award for the first edition of her textbook *Introduction to Cosmology*, and is co-author of the books *Foundations of Astrophysics*, *Interstellar and Intergalactic Medium*, and *Stellar Structure and Evolution*.

Celestial and Stellar Dynamics

Barbara Ryden

The Ohio State University

CAMBRIDGE
UNIVERSITY PRESS

Shaftesbury Road, Cambridge CB2 8EA, United Kingdom

One Liberty Plaza, 20th Floor, New York, NY 10006, USA

477 Williamstown Road, Port Melbourne, VIC 3207, Australia

314–321, 3rd Floor, Plot 3, Splendor Forum, Jasola District Centre,
New Delhi – 110025, India

103 Penang Road, #05–06/07, Visioncrest Commercial, Singapore 238467

Cambridge University Press is part of Cambridge University Press & Assessment,
a department of the University of Cambridge.

We share the University's mission to contribute to society through the pursuit of
education, learning and research at the highest international levels of excellence.

www.cambridge.org
Information on this title: www.cambridge.org/highereducation/isbn/9781108836432

DOI: 10.1017/9781108871013

First published 2025

Printed in the United Kingdom by CPI Group Ltd, Croydon CR0 4YY, 2025

A catalogue record for this publication is available from the British Library

A Cataloging-in-Publication data record for this book is available from the Library of Congress

ISBN 978-1-108-83643-2 Hardback
ISBN 978-1-108-81901-5 Paperback

Additional resources for this publication at www.cambridge.org/osas_csd

For Jim Gunn, who taught me most of what I know about dynamics (and an infinitesimal fraction of what he knows about dynamics).

Contents

Preface

This textbook is part of a series based on the curriculum for astronomy graduate students at The Ohio State University (OSU). In this curriculum, first-year graduate students take a five credit-hour course "Observed Properties of Astronomical Systems." This is followed by six courses, each of two three credit-hours: "Atomic and Radiative Processes in Astrophysics," "Stellar Structure and Evolution," "Dynamics," "Cosmology," "Numerical and Statistical Methods in Astrophysics," and "The Interstellar Medium and the Intergalactic Medium." The philosophy of the OSU graduate program, however, is best encapsulated in the two credit-hour course "Order of Magnitude Astrophysics," which is offered every year to first- and second-year students. In this course, students work together to solve a wide range of astrophysical problems, using basic physical principles to find back-of-envelope solutions.

The Ohio State Astrophysics Series (OSAS), of which this is the third volume, is a projected series of books based on lecture notes for the six core courses and the first-year "Observed Properties" course. These textbooks will not be exhaustive monographs, but will instead adopt the back-of-envelope philosophy of the "Order of Magnitude" course to emphasize the most important physical principles in each subfield of astrophysics. The goal is to make our series a point of entry into the deeper and more detailed classic textbooks in our field. Although each volume in OSAS will stand on its own, care will be taken to unify notation and vocabulary to the extent possible across volumes.

This textbook uses the cgs (centimeter, gram, second) system of units commonly used in graduate education in astronomy. It also uses the most common astronomical distance units: the solar radius ($R_\odot$), the astronomical unit (au), and the parsec (pc). In addition, masses are given in units of the solar mass ($M_\odot$) and luminosities in units of the solar luminosity ($L_\odot$). On small scales, when we examine individual photons and other particles, the electron-volt (eV) will be a useful small unit of energy, with $1\,\mathrm{eV} = 1.602 \times 10^{-12}\,\mathrm{erg}$.

Other helpful conversion factors, and the values of physical and astronomical constants, are included in the Appendix. The Cambridge University Press website for this book (www.cambridge.org/osas_csd) contains ancillary materials, such as a solutions manual and links to Jupyter notebooks in Python for recreating and modifying figures in the textbook.

Plotting of figures was done by Richard Pogge (OSU), in his role as OSAS technical editor. Particular thanks are due to the colleagues who provided their original data: Todd Thompson (Figure 2.10), Carl Murray (Figures 3.5 and 3.7), Ian Wong (Figure 5.7), Rebekah Dawson (Figure 5.8(a)), Thomas de Boer (Figures 8.2 and 10.2(a)), Ana Bonaca (Figure 9.8), and Elizandra Martinazzi (Figure 10.1). Thanks are also due to Annika Peter, who was generous in sharing course materials, and to the "Dynamics" students who offered helpful suggestions for improvement.

Newtonian Dynamics

Of the classical problems of physics, there is one which might be
picked out as simplest to state but slowest to approach solution.
"What is the fate of a self-gravitating system of point masses
interacting according to Newton's laws?"

Jeremiah P. Ostriker (1937–)
Dynamics of Star Clusters, "Some Summary Remarks" [1985]

The noun "dynamics" entered the English language in the eighteenth century, when natural philosophers, following the lead of Isaac Newton, began thinking of motion in terms of applied forces and the resulting accelerations. In 1788, the *New Royal Encyclopaedia* contained the definition, "*Dynamics* is the science of moving powers; more particularly of the motion of bodies that mutually act on one another." This is still a useful definition. For the purposes of this book, we can define dynamics as the study of objects that move while interacting through mutual forces.[1]

More specifically, this book deals with **celestial dynamics** (also known as celestial mechanics) and with **stellar dynamics**. Celestial dynamics is the study of systems containing a relatively small number of objects moving under their mutual gravitational attraction. The study of celestial dynamics began with the study of planets and comets within the solar system, but received renewed interest with the discovery of exoplanetary systems around stars other than the Sun. Applying the techniques of celestial dynamics to a system with N objects, the orbit of each individual object can be computed under the influence of the other $N - 1$ objects. By contrast, stellar dynamics is the study of systems containing a large number of objects moving under their mutual gravitational attraction. When the number N of objects in the system is sufficiently large, it becomes useful to describe the system in terms of its smoothly averaged

---

[1] The word "dynamics," by the way, is linked to the ancient Greek word *dúnamai*, meaning "I am able" or "I am strong enough." Let this be your inspiration if the study of dynamics ever seems difficult.

local properties, such as the mass density $\rho(\vec{r}, t)$, bulk velocity $\vec{u}(\vec{r}, t)$, velocity dispersion $\sigma(\vec{r}, t)$, and so forth. (As we'll see later, the techniques of stellar dynamics have similarities with those of gas dynamics: you don't need to know the trajectory of each air molecule to know which way the wind blows.)

There are some situations where the tools of celestial dynamics are obviously the most useful. Consider, for example, a system containing one star and eight planets. There are only $(9 \cdot 8)/2 = 36$ unique pairs of objects in such a system, and treating the system as a collection of nine point masses interacting with each other is the most useful course. There are also some situations where the tools of stellar dynamics are obviously the most useful. Consider, for example, a galaxy containing 300 billion stars. In this case, there are 4.5×10^{22} pairs of objects. This is an awkwardly large number of gravitational interactions; thus, using averaged properties such as ρ, $\vec{u}$, and σ is the most useful approach in this case.

There are also intermediate cases where the techniques of both celestial dynamics and stellar dynamics can be applied. Consider, for example, a star cluster with 1000 stars. Using numerical techniques, it is feasible to trace all the stellar orbits under the assumption they are point masses interacting gravitationally; this treats the cluster as a jumbo-sized solar system, and involves using the concepts of celestial dynamics. However, if the cluster is nearly spherical, it is useful to compute the average density $\rho(r)$, the potential $\Phi(r)$, and so forth; this treats the cluster as a miniature galaxy, and involves using the concepts of stellar dynamics.

Regardless of the number of individual stars (or other bodies) in a system, we will assume, in the course of this book, that they obey the laws of Newtonian physics. In order to justify that assumption, a brief overview of Newtonian dynamics (and the limitations of Newtonian physics) is in order.

1.1 Newtonian Dynamics

The story of Newtonian dynamics begins with Newton's three laws of motion, as set forward in the *Principia Mathematica* in 1687. In modern language (avoiding the Latin of the *Principia*) the three laws of motion are:

1. In the absence of outside forces, the velocity of an object is constant.
2. The acceleration of an object is directly proportional to the net outside force and inversely proportional to the object's mass.
3. If two objects exert forces on each other, the forces are equal in magnitude but opposite in direction.

The three laws of motion apply to all of Newtonian dynamics. However, celestial and stellar dynamics restrict themselves to the case where the applied forces are purely gravitational. Thus, we must examine Newton's law of gravity, as it applies to a pair of objects.

Consider two objects of mass m_1 and m_2. If $m_1 > m_2$, we call m_1 the **primary** and m_2 the **secondary**. According to Newton's law of gravity, each object feels a gravitational force pulling it toward the other object; the magnitude of the force is

$$F_{\text{grav}} = \frac{Gm_1 m_2}{r^2}, \tag{1.1}$$

where G is Newton's gravitational constant and r is the distance between the objects. Equation 1.1 holds true when the two objects are mathematical points. It also holds true, as Newton was able to demonstrate, when each of the two objects contains a spherically symmetric mass distribution; in this case, r is the distance between the centers of the objects.

Strictly speaking, Einstein's theory of general relativity gives a better prediction of orbital motion; it is needed, for instance, to accurately describe the precession of the planet Mercury's orbit. However, Newtonian gravity gives an extremely accurate approximation in the limit that the curvature of spacetime is very gentle or, equivalently, when the separation between two objects is very large compared to the Schwarzschild radius of the primary, $R_{\text{sch}} = 2Gm_1/c^2$. If masses are given in units of the Sun's mass, $1\,\text{M}_\odot = 1.9884 \times 10^{33}$ g, this requirement can be written as

$$r \gg \frac{2Gm_1}{c^2} \approx 3\,\text{km}\left(\frac{m_1}{1\,\text{M}_\odot}\right). \tag{1.2}$$

Since stars, in general, are much bigger than their Schwarzschild radii, general relativity is important only for very close pairs in which the gravitating objects are neutron stars or black holes.

Newton's second law of motion tells us that an object of mass m experiencing a net force $\vec{F}$ will have an acceleration $\vec{a} = \vec{F}/m$. Thus, in our example of two objects with $m_1 > m_2$, the massive primary experiences the smaller acceleration:

$$a_1 = \frac{F_{\text{grav}}}{m_1} = \frac{Gm_2}{r^2}, \tag{1.3}$$

as compared to

$$a_2 = \frac{Gm_1}{r^2} \tag{1.4}$$

for the less massive secondary. In Newtonian dynamics, there is no theoretical upper limit to the relative speed of two objects moving through space. In Einstein's relativity, by contrast, the upper speed limit for such relative motion is $v = c$, where $c = 299\,792.458\,\text{km s}^{-1}$ is the speed of light through a vacuum.

Can we reasonably expect the speeds produced by gravitational acceleration to be small compared to c? If we assume a secondary of negligible mass is on a circular orbit of radius r, its centripetal acceleration, $a = v_2^2/r$, is equal to the gravitational acceleration when

$$\frac{v_2^2}{r} = \frac{Gm_1}{r^2}, \tag{1.5}$$

or

$$v_2^2 = \frac{Gm_1}{r}. \tag{1.6}$$

For this speed to satisfy $v_2^2 \ll c^2$, we require an orbit of size

$$r \gg \frac{Gm_1}{c^2} \sim R_{\text{sch}}. \tag{1.7}$$

Thus, as long as we focus on stars, planets, and other objects much larger than their Schwarzschild radius, we expect orbital speeds to be smaller than the speed of light. The only case where the relativistic speed limit becomes important is the previously mentioned case of very close pairs containing neutron stars or black holes.

1.2 Two-Body Systems: Conserved Properties

Consider a system of two spherical masses orbiting under their mutual gravitational attraction. The separation between the objects is large enough that purely Newtonian physics can be used. Such a system may seem boringly simple, but there are reasons for studying a two-body system. First, it has analytic solutions for the motion of the two objects. In an era when electronic computers are cheap, this seems like an old-fashioned reason; however, analytic solutions are capable of yielding interesting physical insights. Second, looking at such a simple system tells us about the conservation laws associated with Newtonian dynamics.

Historically, the two-body problem is of great importance as well. Isaac Newton didn't develop his ideas about forces and acceleration simply by staring at falling apples. He was inspired by the laws of planetary motion derived by Johannes Kepler from the observational data of Tycho Brahe. At the accuracy provided by Tycho's naked-eye observations, a system consisting of the Sun plus a single planet could be well approximated as an isolated two-body system. (The small perturbations caused by planets acting on each other were only readily measurable during the telescopic era.) Kepler's three laws of planetary motion, although they are a staple of introductory astronomy classes, were not extracted from Kepler's voluminous writings and elevated as

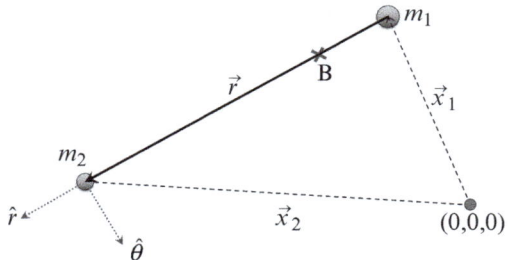

Figure 1.1 A system of two mutually gravitating spheres with masses m_1 and m_2 (where $m_1 > m_2$). The point labeled "B" is the barycenter of the system.

"*THE* Three Laws of Planetary Motion" until the nineteenth century. Today, Kepler's three laws are usually written as:

1. The orbit of a planet is an ellipse with the Sun at one focus.
2. A line segment joining a planet to the Sun sweeps out equal areas in equal times.
3. The square of a planet's orbital period $\mathcal{P}$ is proportional to the cube of the semimajor axis a of its orbit ($\mathcal{P}^2 = \kappa a^3$).

Kepler also showed that the third law applies to the Galilean moons of Jupiter, although with a larger proportionality constant κ. This hinted that his laws of "planetary" motion could be generalized. However, it was Isaac Newton who successfully made the grand leap to universal laws of motion and a universal law of gravity. Let us put Newton's laws to work on a system consisting of two spherical masses.

The geometry of a two-body system is shown in Figure 1.1. The primary has mass m_1, while the secondary has mass $m_2 < m_1$. To trace the motion of the two objects, we set up an **inertial** reference frame. An inertial reference frame is a coordinate system in which Newton's laws of motion hold true. The origin of our inertial reference frame is at an arbitrary location $(0, 0, 0)$. Relative to the origin, the centers of the two massive objects are at positions $\vec{x}_1$ and $\vec{x}_2$. The relative position vector of the two objects is $\vec{r} \equiv \vec{x}_2 - \vec{x}_1$. That is, we define the vector as pointing from the primary to the secondary, as shown in Figure 1.1. For convenience, we label the unit vector in the direction of $\vec{r}$ as $\hat{r} \equiv \vec{r}/r$.

In the inertial reference frame, the acceleration of the primary is (compare to Equation 1.3)

$$\frac{d^2\vec{x}_1}{dt^2} = Gm_2 \frac{\hat{r}}{r^2}. \tag{1.8}$$

The acceleration of the secondary is (Equation 1.4)

$$\frac{d^2\vec{x}_2}{dt^2} = -Gm_1\frac{\hat{r}}{r^2}. \tag{1.9}$$

Subtracting Equation 1.8 from Equation 1.9 yields

$$\frac{d^2}{dt^2}(\vec{x}_2 - \vec{x}_1) = -G(m_1 + m_2)\frac{\hat{r}}{r^2} \tag{1.10}$$

or, more compactly,

$$\frac{d^2\vec{r}}{dt^2} = -\mu\frac{\hat{r}}{r^2}, \tag{1.11}$$

where $\mu \equiv G(m_1 + m_2)$. Equation 1.11 is the **equation of relative motion** for the two objects in the system. This equation reminds us that when we use Newtonian dynamics to estimate the mass of a self-gravitating system, what we find is not the mass m, but the product Gm, where $G = 6.6743 \times 10^{-8}\,\mathrm{cm^3\,g^{-1}\,s^{-2}}$ is the gravitational constant. The value of G is not particularly well known compared to that of other physical constants. Thus, the product $\mu = Gm$, known as the **standard gravitational parameter**, is often known more accurately than G and m are known separately. For example, the orbital motion of planets and other bodies in the solar system tells us that the Sun has $\mu_\odot = GM_\odot = 1.327\,1244 \times 10^{26}\,\mathrm{cm^3\,s^{-2}}$. To avoid blushing with embarrassment at how poorly we know G, we will use the standard gravitational parameter μ rather than the mass m whenever possible. For our binary system, $\mu \equiv G(m_1 + m_2)$, with $\mu_1 = Gm_1$ and $\mu_2 = Gm_2$ for the individual components.

The equation of relative motion (Equation 1.11) can also be written as

$$\frac{d^2\vec{r}}{dt^2} = -\frac{d\Phi_K}{dr}\hat{r}, \tag{1.12}$$

where

$$\Phi_K(r) = -\frac{\mu}{r} \tag{1.13}$$

is the **Keplerian potential**.[2] Here, we are using the common (but not universal) sign convention that Φ_K is negative when r is small, and the usual normalization that $\Phi_K \to 0$ as $r \to \infty$.

Taking the cross product of Equation 1.11 with the relative position vector $\vec{r}$, we find

$$\vec{r} \times \frac{d^2\vec{r}}{dt^2} = 0 \tag{1.14}$$

---

[2] Although the concept of a gravitational potential was not introduced until over a century after Johannes Kepler's death, the term "Keplerian potential" is a posthumous tribute to Kepler's studies of planetary motion.

for motion dictated by Newtonian gravity. Equation 1.14 is equivalent to the equation

$$\vec{r} \times \frac{d\vec{r}}{dt} = \vec{j}, \tag{1.15}$$

where $\vec{j}$ is a constant of integration. (You can verify this assertion by differentiating Equation 1.15 with respect to t and recovering Equation 1.14.)

The vector $\vec{j}$ is related to the angular momentum of the two-body system. The total angular momentum of the system is found using the positions and velocities of the two objects relative to the **barycenter**, or center of mass, of the system. The barycenter (labeled "B" in Figure 1.1) is located on the line segment connecting the centers of the two objects. The position of the primary relative to the barycenter is

$$\vec{r}_1 = -\frac{m_2}{m_1 + m_2} \vec{r}. \tag{1.16}$$

The secondary is farther from the barycenter; its position relative to the barycenter is

$$\vec{r}_2 = \frac{m_1}{m_1 + m_2} \vec{r}. \tag{1.17}$$

Given the velocity of the secondary relative to the primary is $\vec{v} \equiv d\vec{r}/dt$, the velocity of the primary relative to the barycenter is

$$\vec{v}_1 = -\frac{m_2}{m_1 + m_2} \vec{v}. \tag{1.18}$$

The secondary is moving more rapidly relative to the barycenter, with velocity

$$\vec{v}_2 = \frac{m_1}{m_1 + m_2} \vec{v}. \tag{1.19}$$

Using the barycenter as our reference point, the orbital angular momentum of the two-body system is

$$\vec{J}_{\text{orb}} = m_1(\vec{r}_1 \times \vec{v}_1) + m_2(\vec{r}_2 \times \vec{v}_2). \tag{1.20}$$

Substituting for $\vec{r}_1, \vec{r}_2, \vec{v}_1,$ and $\vec{v}_2$, we find

$$\vec{J}_{\text{orb}} = m_r(\vec{r} \times \vec{v}), \tag{1.21}$$

where

$$m_r \equiv \frac{m_1 m_2}{m_1 + m_2} \tag{1.22}$$

is the **reduced mass** of the two-body system. Thus, the constant of integration $\vec{j}$ in Equation 1.15 is

$$\vec{j} = \frac{\vec{J}_{\text{orb}}}{m_r}, \tag{1.23}$$

showing that Equation 1.15 is a statement of angular momentum conservation. Deriving this result didn't require knowing that gravitational force follows an inverse square law. It simply required that the mutual force between m_1 and m_2 be parallel to the vector $\vec{r}$ and that the magnitude of the force be a function only of the separation r; in the language of classical physics, such a force is called a **central force**.

Since $\vec{j}$ is constant, the motion of the secondary relative to the primary is confined to a fixed plane perpendicular to $\vec{j}$; we call this plane the **orbital plane**. To describe the relative motion of the two objects in this plane, it is useful to use polar coordinates; the radial coordinate $r = |\vec{r}|$ is measured relative to the primary, and the angular coordinate θ increases in the direction of the secondary's motion relative to the primary. In polar coordinates, velocity in the plane is expressed as

$$\frac{d\vec{r}}{dt} = \frac{dr}{dt}\hat{r} + r\frac{d\theta}{dt}\hat{\theta}. \tag{1.24}$$

Making use of Equation 1.15, this implies

$$j = r^2\frac{d\theta}{dt} = \text{constant}. \tag{1.25}$$

During a brief time interval dt, the vector $\vec{r}$ connecting the secondary and primary moves through a small angle $d\theta$, sweeping out a long skinny triangle with area

$$dA = \frac{1}{2}r(rd\theta). \tag{1.26}$$

Combining Equations 1.25 and 1.26 yields Kepler's second law of planetary motion:

$$\frac{dA}{dt} = \frac{1}{2}r^2\frac{d\theta}{dt} = \frac{j}{2} = \text{constant}. \tag{1.27}$$

Kepler's second law is thus equivalent to saying that the specific angular momentum j is constant.

Previously, we took the cross product of the equation of relative motion (Equation 1.11) with the relative position $\vec{r}$. Now, let's try something different: let's take taking the dot product of Equation 1.11 with the relative velocity $d\vec{r}/dt$. This yields

$$\frac{d\vec{r}}{dt} \cdot \frac{d^2\vec{r}}{dt^2} + \mu\frac{d\vec{r}}{dt} \cdot \frac{\hat{r}}{r^2} = 0. \tag{1.28}$$

Integrating Equation 1.28 over time, we find

$$\frac{1}{2}\frac{d\vec{r}}{dt} \cdot \frac{d\vec{r}}{dt} - \frac{\mu}{r} = \frac{1}{2}v^2 - \frac{\mu}{r} = \epsilon, \tag{1.29}$$

where ϵ is a constant of integration.

The scalar ϵ is related to the **orbital energy** of the two-body system. The orbital energy is found by adding together the gravitational potential energy of the system and the kinetic energy of the two objects as they move on their orbits. If velocities are taken relative to the center of mass, the orbital energy is

$$E_{\text{orb}} = \frac{1}{2}m_1 v_1^2 + \frac{1}{2}m_2 v_2^2 - \frac{Gm_1 m_2}{r}. \tag{1.30}$$

Using the values of v_1 and v_2 given in Equations 1.18 and 1.19, we find that

$$E_{\text{orb}} = m_r \left(\frac{1}{2}v^2 - \frac{\mu}{r} \right), \tag{1.31}$$

where m_r is the reduced mass and μ is the standard gravitational parameter. Thus, the constant of integration ϵ in Equation 1.29 is

$$\epsilon = \frac{E_{\text{orb}}}{m_r}, \tag{1.32}$$

showing that Equation 1.29 is a statement of energy conservation; the sum of the kinetic energy and the gravitational potential energy is constant in a two-body system.

We have found four conserved properties as the primary and secondary move relative to each other. These are the three components of the specific angular momentum $\vec{j}$ and the specific energy ϵ, or equivalently the three components of the orbital angular momentum $\vec{J}_{\text{orb}}$ and the orbital energy E_{orb}.

1.3 Two-Body Systems: Orbits

The equation of relative motion,

$$\frac{d^2 \vec{r}}{dt^2} = -\mu \frac{\hat{r}}{r^2}, \tag{1.33}$$

is a second-order differential equation. If we know the relative position $\vec{r}_0$ and the relative velocity $\vec{v}_0$ of the two objects at an initial time t_0, and if the mass μ is fixed, we will able to compute $\vec{r}(t)$ at any later time – or any earlier time, for that matter.

In our chosen polar coordinates, the relative velocity is given by Equation 1.24, and the relative acceleration is

$$\frac{d^2 \vec{r}}{dt^2} = \left[\frac{d^2 r}{dt^2} - r \left(\frac{d\theta}{dt} \right)^2 \right] \hat{r} + \left[\frac{1}{r} \frac{d}{dt} \left(r^2 \frac{d\theta}{dt} \right) \right] \hat{\theta}. \tag{1.34}$$

However, if the acceleration results from a central force (such as Newtonian gravity), Equation 1.25 tells us that $r^2(d\theta/dt) = j =$ constant. Thus,

$$\frac{d^2\vec{r}}{dt^2} = \left[\frac{d^2r}{dt^2} - r\left(\frac{d\theta}{dt}\right)^2\right]\hat{r} = \left[\frac{d^2r}{dt^2} - \frac{j^2}{r^3}\right]\hat{r}. \tag{1.35}$$

Combining Equation 1.33 with Equation 1.35 gives us the equation of relative motion in scalar form:

$$\frac{d^2r}{dt^2} - \frac{j^2}{r^3} = -\frac{\mu}{r^2}. \tag{1.36}$$

If we solve this equation for $r(t)$, then $\theta(t)$ follows from the requirement that $d\theta/dt = j/r^2$. Our astronomical ancestors were annoyed by the fact that Equation 1.36 doesn't have a general analytic solution. However, they realized that they could make the substitution $u \equiv 1/r$, then use the relation $d\theta = ju^2dt$ to rewrite Equation 1.36 as

$$\frac{d^2u}{d\theta^2} + u = \frac{\mu}{j^2}. \tag{1.37}$$

Since μ/j^2 is constant, Equation 1.37 has one obvious solution:

$$u = \frac{\mu}{j^2}. \tag{1.38}$$

This means it is possible, with the right initial conditions, for the secondary to be on a circular orbit relative to the primary; in this case, the orbital radius $r = j^2/\mu$ is dictated by the mass of the binary system and its specific angular momentum j.

The most general solution to Equation 1.37 is

$$u(\theta) = \frac{\mu}{j^2}\left[1 + e\cos(\theta - \theta_0)\right], \tag{1.39}$$

where the phase θ_0 and dimensionless amplitude e of the sinusoidal component are constants of integration, whose values are chosen to match the initial conditions. The convention in celestial dynamics is that e must be a non-negative number; given the trigonometric identity $\cos\alpha = -\cos(\alpha \pm \pi)$, we are free to flip the sign of e as long as we simultaneously shift the phase by π radians. The distance between the primary and secondary, as a function of θ, is thus

$$r(\theta) = \frac{1}{u(\theta)} = \frac{j^2/\mu}{1 + e\cos(\theta - \theta_0)}. \tag{1.40}$$

This is the equation for a **conic section**. If $e < 1$, the orbit is bound, with r always having a finite value. In the special case $e = 0$, the orbit is circular; if $0 < e < 1$, it is an ellipse. If $e \geq 1$, the orbit is unbound. In the special case $e = 1$, the orbit is a parabola; if $e > 1$, it is a hyperbola.

The standard form for a conic section in polar coordinates is

$$r(\theta) = \frac{a(1 - e^2)}{1 + e \cos(\theta - \theta_0)}, \tag{1.41}$$

where e is the eccentricity and a, in the case of an elliptical orbit, is the semi-major axis of the ellipse. Equation 1.41 also describes hyperbolic orbits; this is useful, for instance, in describing the orbit of interstellar comets as they pass through the inner solar system. However, since hyperbolas have $e > 1$, Equation 1.41 requires a negative value for a when the orbit is unbound. The closest approach between the primary and secondary on a hyperbolic orbit occurs when $\theta = \theta_0$, with minimum separation $r_{min} = |a|(e^2 - 1)$. If the primary and secondary start at a very large distance from each other, with $r \gg r_{min}$, the value of θ goes from $\cos(\theta - \theta_0) = -1/e$ initially, through $\cos(\theta - \theta_0) = 0$ at closest approach, to $\cos(\theta - \theta_0) = -1/e$ as $r \to \infty$ again. Keeping in mind that $\theta - \theta_0$ is initially negative but ends up positive, the total angle through which the relative motion is deflected works out to be

$$\Delta\theta = \pi - 2\cos^{-1}(1/e) = 2\sin^{-1}(1/e) \qquad [e \geq 1]. \tag{1.42}$$

Thus, in the parabolic limit ($e = 1$), $\Delta\theta = \pi$ and the primary and secondary make U-turns. In the limit $e \to \infty$, the deflection becomes negligible, with $\Delta\theta \approx 2/e \to 0$.

Comparing Equations 1.40 and 1.41, we find that the orbital parameters a and e are related to the mass and specific energy of the binary system, with

$$a(1 - e^2) = \frac{j^2}{\mu}. \tag{1.43}$$

Looking more specifically at bound orbits, with $e < 1$ and $a > 0$, we find that a circular orbit ($e = 0$) is the orbit that minimizes a for a given mass μ and specific angular momentum j.

Figure 1.2 shows the motion of the secondary relative to the primary along a fairly eccentric ($e = 0.8$) elliptical orbit. Notice that the primary is at one focus of the ellipse; our two-body system follows Kepler's first law of planetary motion if we take the primary to be the Sun and the secondary to be a planet. The constant θ_0 is called the **argument of periapsis**. When $\theta = \theta_0$, the secondary comes as close as it can get to the primary. The point of closest approach is called the **periapsis**, with $r_{pe} = a(1 - e)$. The point at which the secondary is farthest from the primary, when $\theta = \theta_0 \pm \pi$, is the **apoapsis**, with $r_{ap} = a(1 + e)$. The line connecting the periapsis and apoapsis, passing through both foci of the ellipse, is called the **line of apsides**.[3]

---

[3] The word "apsis," with plural form "apsides," comes from an ancient Greek word meaning a wheel or arch. It is cognate with the architectural term "apse," referring to the rounded end of a cathedral.

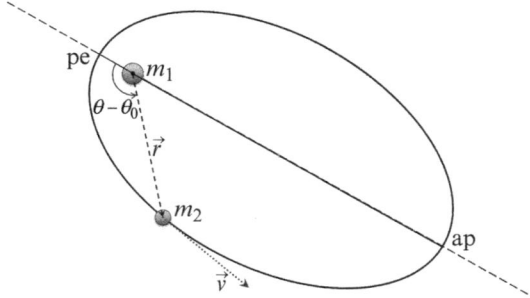

Figure 1.2 Motion of the secondary (mass m_2) relative to the primary (mass m_1) on a bound elliptical orbit with $e = 0.8$. The periapsis (labeled "pe") and apoapsis (labeled "ap") define the line of apsides.

Usually, the angular coordinate θ is measured from a zero point that is chosen for convenience. As an example, for an object orbiting the Sun, the general convention is that $\theta = 0$ at the point where the orbiting object passes through the ecliptic plane going from south to north.[4] In addition, it is often useful to define an alternate angular coordinate v whose zero point is at the periapsis of the object's orbit. This coordinate, called the **true anomaly**, is defined as $v(t) \equiv \theta(t) - \theta_0$. In terms of the true anomaly v, an elliptical orbit can be written more simply as

$$r(v) = \frac{a(1 - e^2)}{1 + e \cos v}.$$ (1.44)

Although there is no general analytic relation for the true anomaly as a function of time, for mildly eccentric orbits there is a useful expansion in powers of e. If we define the **mean motion** of an orbit as $n \equiv 2\pi/\mathcal{P}$, then $v(t)$ can be approximated as

$$v(t) \approx nt + 2e \sin(nt) + \frac{5}{4}e^2 \sin(2nt)$$ (1.45)

if we set $t = 0$ at an instant when the orbiting object is at periapsis. Using this approximation for $v(t)$ in Equation 1.44, we derive a useful approximation for $r(t)$:

$$r(t) \approx a\left(1 - e \cos(nt) - \frac{1}{2}e^2[\cos(2nt) - 1]\right).$$ (1.46)

Consider an orbiting object going from periapsis at $v = nt = 0$ to apoapsis at $v = nt = \pi$. During this time interval, the distance r continuously increases from $r_{\mathrm{pe}} = a(1 - e)$ to $r_{\mathrm{ap}} = a(1 + e)$. In the mildly eccentric limit, the moment

---

[4] The ecliptic plane, defined by the Earth's orbit around the Sun, divides the entire universe in half: the "south" half is the one containing the Earth's south pole. This is a shamelessly Earth-centered convention, chosen for the convenience of Earth-based astronomers.

when $r = a$ corresponds to $nt \approx \pi/2 - e$ and $v \approx \pi/2 + e$. The Earth's orbit, for example, has $\mathcal{P} = 365.256\,\text{d}$ and $e = 0.0167$. This implies that the moment when the Earth is exactly 1 au from the Sun (and moving away) is not $t = \mathcal{P}/4 = 91.31\,\text{d}$ after periapsis; instead, it comes earlier by an amount $\delta t \approx -e\mathcal{P}/2\pi \approx -0.97\,\text{d}$.

On an elliptical orbit, the periapsis and apoapsis are the only points where $\vec{v}$ is perpendicular to $\vec{r}$; this means that at these two points,

$$j \equiv |\vec{r} \times \vec{v}| = rv. \tag{1.47}$$

Since, from Equation 1.43,

$$j = \mu^{1/2}a^{1/2}(1 - e^2)^{1/2}, \tag{1.48}$$

the speed of the secondary at periapsis must be

$$v_{\text{pe}} = \frac{\mu^{1/2}a^{1/2}(1 - e^2)^{1/2}}{a(1 - e)} = \frac{\mu^{1/2}}{a^{1/2}}\left(\frac{1 + e}{1 - e}\right)^{1/2}. \tag{1.49}$$

The specific energy calculated at periapsis is then

$$\epsilon = \frac{1}{2}v_{\text{pe}}^2 - \frac{\mu}{r_{\text{pe}}} = \frac{\mu(1 + e)}{2a(1 - e)} - \frac{\mu}{a(1 - e)} = -\frac{\mu}{2a}. \tag{1.50}$$

Since $\epsilon = -\mu/2a$ is constant along the orbit, we can use the relation

$$\frac{1}{2}v(r)^2 - \frac{\mu}{r} = -\frac{\mu}{2a} \tag{1.51}$$

to write down a simple relation for $v(r)$:

$$v(r) = \mu^{1/2}\left(\frac{2}{r} - \frac{1}{a}\right)^{1/2}. \tag{1.52}$$

This relation is known as the *vis viva* equation. Using the mildly eccentric approximation for $r(t)$, as given in Equation 1.46, we can make the expansion

$$v(t) \approx \left(\frac{\mu}{a}\right)^{1/2}\left(1 + e\cos(nt) + \frac{1}{4}e^2[3\cos(2nt) - 1]\right). \tag{1.53}$$

Note also that $\epsilon = -\mu/2a$ corresponds to a conserved orbital energy of

$$E_{\text{orb}} = m_r\epsilon = -\frac{m_1 m_2}{m_1 + m_2}\frac{G(m_1 + m_2)}{2a} = -\frac{Gm_1 m_2}{2a}. \tag{1.54}$$

As the secondary swoops around on an elliptical orbit relative to the primary, it takes a time $\mathcal{P}$ for the true anomaly v to go through 2π radians. During this time, the vector $\vec{r}$ sweeps out an area A equal to the area enclosed by the ellipse. That is, $A = \pi ab = \pi a^2(1 - e^2)^{1/2}$, where $b = a(1 - e^2)^{1/2}$ is the semiminor axis of the ellipse. However, Kepler's second law, as given in Equation 1.27, tells us that

$$A = \int_0^{\mathcal{P}} \frac{dA}{dt}\, dt = \int_0^{\mathcal{P}} \frac{j}{2}\, dt = \frac{j}{2}\mathcal{P}. \tag{1.55}$$

Thus, using the geometrical relation for the area enclosed by the ellipse, we find

$$\pi a^2 (1 - e^2)^{1/2} = \frac{j}{2}\mathcal{P}. \tag{1.56}$$

Substituting the value of j from Equation 1.48, this becomes

$$\pi a^2 (1 - e^2)^{1/2} = \frac{1}{2}\left[\mu a(1 - e^2)\right]^{1/2}\mathcal{P}. \tag{1.57}$$

The factors involving $1 - e^2$ cancel on either side, and we are left with Kepler's third law of planetary motion:

$$\frac{2\pi}{\mu^{1/2}} a^{3/2} = \mathcal{P}. \tag{1.58}$$

In terms of the mean motion $n = 2\pi/\mathcal{P}$, this can be written as

$$n = \frac{\mu^{1/2}}{a^{3/2}}. \tag{1.59}$$

So far, we have considered the motion of the secondary relative to the primary. Within the solar system, for instance, Kepler's third law relates the average distance from the Sun to a planet (a) to the planet's orbital period around the Sun ($\mathcal{P}$). You might argue, "If it's good enough for Kepler, it's good enough for me!" However, there are circumstances when it is more useful to measure the motion of both the secondary and the primary relative to the barycenter of the system. If there are no external forces acting on the binary system, then the barycenter moves at constant velocity, and provides us with a ready-made inertial reference frame.

Figure 1.3 shows the same binary system showcased in Figure 1.2, with the added assumption that $m_1 = 2m_2$. The orbits of both the primary and secondary are shown relative to the barycenter. Each of the two objects moves on an elliptical orbit with the same eccentricity and line of apsides as their relative orbit. The barycenter of the system is at one focus of each object's elliptical orbit. The primary's orbit about the barycenter has semimajor axis

$$a_1 = \frac{m_2}{m_1 + m_2} a, \tag{1.60}$$

while the secondary has a larger orbit, with

$$a_2 = \frac{m_1}{m_1 + m_2} a. \tag{1.61}$$

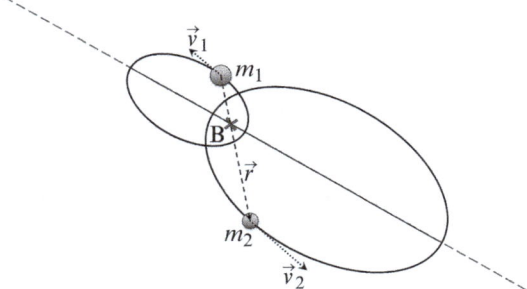

Figure 1.3 Motion of the primary (mass m_1) and the secondary (mass $m_2 = m_1/2$) relative to the barycenter (B). The orbits have eccentricity $e = 0.8$.

The ratio of the size of the orbits is $a_1/a_2 = m_2/m_1$. Equations 1.18 and 1.19 remind us that this is also the ratio of the velocities of the two objects relative to the barycenter. Thus, the orbital period $\mathcal{P}$ is the same regardless of whether you track the motion of the primary relative to the barycenter, the secondary relative to the barycenter, or the secondary relative to the primary.

For some binary systems of interest, the secondary is much less massive than the secondary. For instance, the ratio of the Earth's mass to the Sun's mass is $m_2/m_1 = 3.003 \times 10^{-6}$; given the average Sun–Earth distance of 1.496×10^8 km, this implies that the Sun–Earth barycenter is only 449 km away from the Sun's central point. Since the Sun's radius is $1\,R_\odot = 695\,700$ km, this means the Sun–Earth barycenter is buried deep in the Sun's core. In this example, the motion of the Earth relative to the Sun's center is nearly indistinguishable from its motion relative to the Sun–Earth barycenter.

However, in many binary star systems the two stars are similar in mass to each other, and the barycenter is nearly midway between the primary and secondary. Even for a system consisting of a star and an exoplanet, with a mass ratio $m_2/m_1 \sim 10^{-3}$ or less, the motion of the massive star can be large enough to be measured. Section 5.1 goes into more detail how the radial velocity of a star can be analyzed to reveal the existence of exoplanets and tell us something about their properties.

Exercises

1.1 Approximate the Earth's orbit relative to the Sun as a circle with a radius of one astronomical unit and an orbital period of one sidereal year. Now imagine that the Sun instantaneously loses a fraction f of its total mass (perhaps through the meddling of technologically advanced, but ethically dubious, space aliens).

(a) Compute the values of a and e for the Earth's new orbit around the lower-mass Sun. [*Hint:* Use the position and velocity of the Earth at the instant of the Sun's mass loss as initial conditions.]

(b) For what values of f will the Earth's new orbit be unbound?

1.2 Consider a system consisting of two point masses, the primary having mass m_1 and the secondary mass $m_2 < m_1$. The masses are attracted to each other by a central force F whose amplitude is given by the law

$$F = Hm_1m_2r, \tag{1.62}$$

where H is a constant with units $g^{-1}\,s^{-2}$.

(a) Show that the orbit of the secondary relative to the primary is an ellipse with the primary at the center.

(b) What is the dependence of the secondary's orbital period $\mathcal{P}$ on the semimajor axis a of its orbit?

1.3 Consider a *different* system consisting of two point masses with $m_1 > m_2$. These masses are attracted to each other by a central force F whose amplitude is given by the inverse cube law

$$F = \frac{Km_1m_2}{r^3}, \tag{1.63}$$

where K is a constant with units $cm^4\,g^{-1}\,s^{-2}$. This potential does not, in general, have a neat analytic form for orbital shapes. However, we can look at limiting cases.

(a) Suppose the point masses start at rest relative to each other, with a separation r. How long will it be until the attractive force causes the masses to meet? [*Hint:* A numerical solution of a differential equation may be useful here.]

(b) Consider the special case where the relative orbit of the masses is a perfect circle with radius r. What is the dependence of the orbital period $\mathcal{P}$ on the orbital radius r?

2

Three-Body Systems

One's too few, three too many.

John Ray (1627–1705)
A Collection of English Proverbs, 2nd edition [1678], p. 342

Echoing the Somerset proverb quoted by John Ray, we can state that if we have only one celestial body, that is too few objects to study celestial dynamics. A system with two bodies, as we saw in the previous chapter, contains interesting physics. But what about a three-body system? We might conclude, considering the possible complexity of such a system, that it contains too many objects to be tractable. However, by imposing some restrictions, we will be able to extract interesting physical insight into systems with $N > 2$.

Before we tackle the three-body problem, let's introduce some useful vocabulary. A gravitationally bound system containing two stars is a **binary system**, from the Latin word *bini*, meaning "two at a time." By extension, a bound system containing three stars would be a ternary system, from the Latin *terni*, meaning "three at a time." However, we follow the usual convention and refer to a system with three stars as a **triple system**. (Since the individual stars in a triple system are called the primary, secondary, and tertiary, this avoids confusion between the similar words "ternary" and "tertiary.")

2.1 Restricted Three-Body Problem

The general three-body problem, with arbitrary initial conditions with bodies of arbitrary relative mass, can be solved only numerically. However, as Leonhard Euler and Joseph-Louis Lagrange found in the eighteenth century, we can gain insight by looking at a special case called the **restricted three-body problem**.

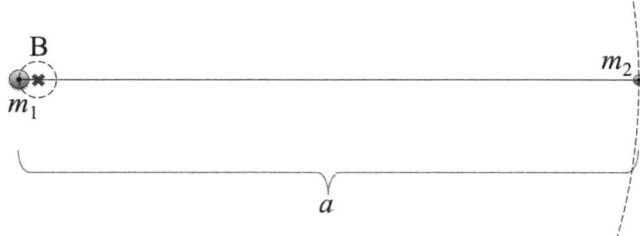

Figure 2.1 Two massive bodies on circular orbits about their barycenter. The secondary contains a fraction $f = m_2/(m_1 + m_2) = 0.03$ of the total mass.

Consider two spherical objects, with mass m_1 and m_2, on circular orbits about their barycenter, as shown in Figure 2.1; the circularity of the orbits is the first restriction in the restricted three-body problem. The distance between m_1 and m_2 is equal to a. The orbital period of the two objects is thus, from Equation 1.58,

$$\mathcal{P} = \frac{2\pi}{\mu^{1/2}} a^{3/2}, \tag{2.1}$$

where $\mu = G(m_1 + m_2)$. The orbital **angular speed** Ω of each object, given the requirement of circular motion, is equal to the mean motion n (Equation 1.59):

$$\Omega = \frac{2\pi}{\mathcal{P}} = \frac{\mu^{1/2}}{a^{3/2}}. \tag{2.2}$$

To this simple binary system we add a third object, with mass $m_3 \ll m_2 \leq m_1$. The third body, or **tertiary**, is merely a test mass, with an unmeasurably tiny gravitational effect on m_1 and m_2; the infinitesimal mass of the tertiary is the next restriction in the restricted three-body problem. We assume that the tertiary moves in the orbital plane defined by the relative motion of m_1 and m_2; the confinement of all motion to a plane is the final restriction in the restricted three-body problem.

Our severe restrictions mean that the idealized three-body system doesn't correspond exactly to any triple systems in the real universe. However, it did provide a useful approximation for early studies of the Sun–Earth–Moon system. In addition, the discovery of the trojan asteroids of Jupiter in the twentieth century provided a new impetus for studies of the restricted three-body problem. In this case, the primary is the Sun, the secondary is Jupiter (with $m_2 = 9.55 \times 10^{-4} m_1$), and the test-mass tertiary is an asteroid (with $m_3 < 10^{-8} m_2$).

Problems in celestial dynamics (and other fields) are often made easier by a sensible choice of coordinates. For the restricted three-body problem, we choose a Cartesian coordinate system (X, Y, Z) whose origin is at the barycenter of the system; since m_3 is infinitesimal, this is the same as the barycenter of the $m_1 + m_2$ binary. We define the Z axis to be perpendicular to the orbital plane, and set the coordinate system rotating about the Z axis with angular speed $\Omega = 2\pi/\mathcal{P}$ in the same sense as the orbital motion of the $m_1 + m_2$ binary. This means that the position of the primary and secondary are constant in the rotating frame of reference. For simplicity, we choose the X axis so that the location of the primary in the rotating frame is

$$(X_1, Y_1) = (-fa, 0), \tag{2.3}$$

where $f \equiv m_2/(m_1 + m_2) \le 1/2$ is the fraction of the total mass possessed by the secondary. The position of the secondary in the rotating frame is

$$(X_2, Y_2) = ([1 - f]a, 0). \tag{2.4}$$

We place our test mass somewhere in the orbital plane, at position (X, Y) as measured in the rotating frame of reference. Its velocity in the Z direction is zero, but in general it has a non-zero velocity in the (X, Y) plane. The distance of the test mass from the primary is

$$r_1^2 = (X + fa)^2 + Y^2. \tag{2.5}$$

Its distance from the secondary is

$$r_2^2 = (X - [1 - f]a)^2 + Y^2. \tag{2.6}$$

The equation of motion for the third-body test mass, using positions, velocities, and accelerations measured in the rotating frame of reference, is *not* given by Newton's second law. A rotating frame is not an inertial frame. The equation of motion in a frame rotating with constant angular speed Ω is

$$\frac{d^2 X}{dt^2} - 2\Omega \frac{dY}{dt} = -\frac{\partial \Phi_{\text{eff}}}{\partial X}, \tag{2.7}$$

$$\frac{d^2 Y}{dt^2} + 2\Omega \frac{dX}{dt} = -\frac{\partial \Phi_{\text{eff}}}{\partial Y}, \tag{2.8}$$

where Φ_{eff} is the **effective potential**,

$$\Phi_{\text{eff}}(X, Y) = -\frac{\Omega^2}{2}(X^2 + Y^2) - \frac{(1 - f)\mu}{r_1} - \frac{f\mu}{r_2}. \tag{2.9}$$

In addition to the Keplerian potentials of the primary and secondary, the effective potential also contains a term proportional to Ω^2; this is the **centrifugal term** of the potential. The terms proportional to dX/dt in Equation 2.7 and dY/dt in Equation 2.8 are the **Coriolis terms**.

In the two-body problem we found four constants of motion: the three components of the specific angular momentum $\vec{j}$ plus the specific energy ϵ.

In the restricted three-body problem, these quantities are still conserved for the massive binary; since the third body has an infinitesimal mass, it does not affect the binary system. However, can we find any constants of motion for the test mass as it is tugged by the two massive objects? Let's start by multiplying Equation 2.7 by dX/dt and Equation 2.8 by dY/dt, then adding. This yields

$$\frac{d^2 X}{dt^2}\frac{dX}{dt} + \frac{d^2 Y}{dt^2}\frac{dY}{dt} = -\frac{\partial \Phi_{\text{eff}}}{\partial X}\frac{dX}{dt} - \frac{\partial \Phi_{\text{eff}}}{\partial Y}\frac{dY}{dt}. \tag{2.10}$$

However, since Φ_{eff} is a function only of X and Y, with no explicit dependence on t, we can write Equation 2.10 more compactly as

$$\frac{d^2 X}{dt^2}\frac{dX}{dt} + \frac{d^2 Y}{dt^2}\frac{dY}{dt} = -\frac{d\Phi_{\text{eff}}}{dt}. \tag{2.11}$$

Integrating over time, this becomes

$$\frac{1}{2}\left[\left(\frac{dX}{dt}\right)^2 + \left(\frac{dY}{dt}\right)^2\right] = -\Phi_{\text{eff}}(X, Y) - \frac{1}{2}C_J, \tag{2.12}$$

where C_J is a constant of integration known as the **Jacobi integral**.[1] Thus, for the motion of the test mass, the conserved property is

$$C_J \equiv -V^2 - 2\Phi_{\text{eff}}, \tag{2.13}$$

where V is the speed of the test mass as measured in the rotating reference frame. Using the effective potential from Equation 2.9, we can write the Jacobi integral more explicitly as

$$C_J = -V^2 + \Omega^2(X^2 + Y^2) + 2\mu\left(\frac{1-f}{r_1} + \frac{f}{r_2}\right). \tag{2.14}$$

The Jacobi integral is useful partly because it tells us which regions in the plane are inaccessible to the test-mass tertiary. If the test mass has a Jacobi integral C_J, then it is excluded from regions where $-2\Phi_{\text{eff}} < C_J$. In those regions, we learn from Equation 2.13, V^2 would have to be negative, and an imaginary speed is not permitted by Newtonian dynamics. Suppose that a test mass has a Jacobi integral C_J. In the orbital plane defined by the massive binary, we can identify a **zero-velocity curve** where $V^2 = 0$ and the test mass is at rest in the rotating frame; from Equation 2.13, the zero-velocity curve is an equipotential curve for $\Phi_{\text{eff}}(X, Y)$. On a zero-velocity curve,

$$C_J = -2\Phi_{\text{eff}}(X, Y) = \Omega^2\left(X^2 + Y^2\right) + 2\mu\left(\frac{1-f}{r_1} + \frac{f}{r_2}\right). \tag{2.15}$$

---

[1] The Jacobi integral is named after Carl Jacobi, the mathematician who gave his name to Jacobian matrices and determinants.

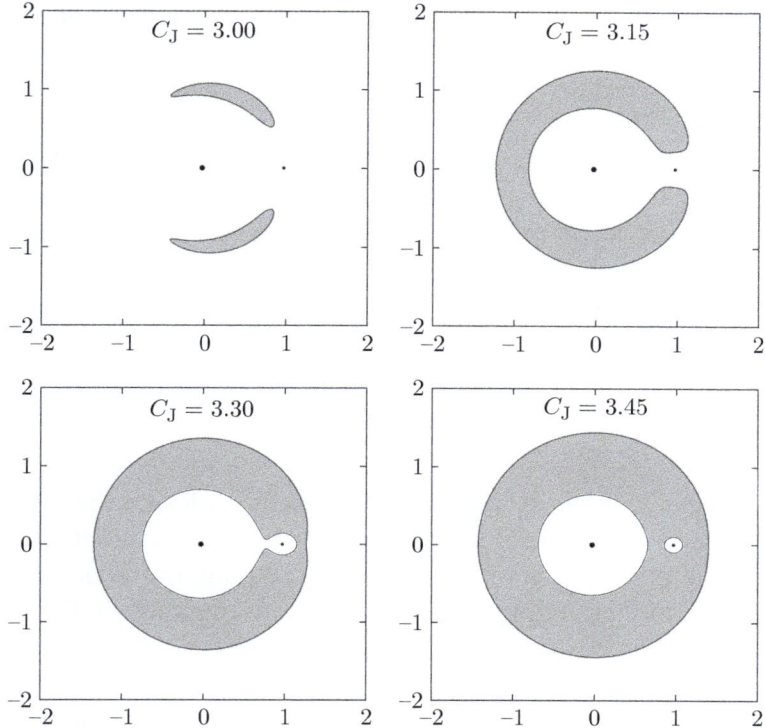

Figure 2.2 Zero-velocity curves for different values of the Jacobi integral C_J in a restricted three-body system with $f = 0.03$. The units are chosen such that $\mu = 1$, $a = 1$, and thus $\Omega = 1$. The barycenter is at $(0, 0)$, the primary is at $(-0.03, 0)$, and the secondary is at $(0.97, 0)$.

However, since Equation 2.2 tells us that $\Omega^2 = \mu/a^3$ for the massive binary, we can rewrite this as

$$C_J = \frac{\mu}{a}\left[\frac{X^2 + Y^2}{a^2} + 2\left(\frac{(1-f)a}{r_1} + \frac{fa}{r_2}\right)\right]. \tag{2.16}$$

Figure 2.2 shows some zero-velocity curves for the $f = 0.03$ system illustrated in Figure 2.1. The curves are the boundaries between the white and gray regions, with the gray regions being the forbidden area where V^2 would have to be negative for the specified C_J. If a test mass is at the boundary between a gray and white region in Figure 2.2, at rest in the rotating frame, then it can move into the white region but not the gray region.

A test mass with a large value of the Jacobi integral, $C_J \gg 3\mu/a$, must be in one of three regions. First, it may be close to the primary, with $r_1 < 2(1 - f)\mu/C_J$; in this region, the test mass and primary approximate an isolated two-body system, with the secondary providing a relatively small perturbing force.

Second, the test mass may be close to the secondary, with $r_2 < 2f\mu/C_J$; in this region, the test mass and secondary approximate a two-body system, with the primary being the perturber. Third, the test mass may be far from the central binary, with $X^2 + Y^2 > C_J/\Omega^2$; in this region, the test mass and the close central binary approximate a two-body system, with the quadrupole moment of the close binary being the perturber. By contrast, a test mass with a sufficiently small value of the Jacobi integral is permitted to wander throughout space while still conserving C_J. However, these particles with low C_J are the celestial equivalent of the *Flying Dutchman*, condemned to have $V > 0$ and never, even instantaneously, coming to a halt in the rotating frame.

2.2 Lagrange Points

In the restricted three-body problem, it is useful to locate the equilibrium points for the test mass. In this case, "equilibrium" means that when a test mass is placed at such a point, at rest in the rotating frame of reference, it remains at rest in the rotating frame. The equilibrium points in the restricted three-body problem are known as the **Lagrange points**, or Lagrangian points. There are five such Lagrange points,[2] as shown in Figure 2.3. The **collinear** Lagrange points, conventionally labeled L_1, L_2, and L_3, lie along the line defined by the primary and secondary. The two remaining Lagrange points, labeled L_4 and L_5, form equilateral triangles with the primary and secondary. The usual convention is to assign the label L_4 to the Lagrange point that leads the secondary, and L_5 to the Lagrange point that trails behind.

Carefully place a test mass at a Lagrange point, making sure it is at rest in the rotating frame, with $dX/dt = dY/dt = 0$. To remain at rest, it must have zero acceleration in the rotating frame, with $d^2X/dt^2 = d^2Y/dt^2$. This means that the equation of motion (Equations 2.7 and 2.8) reduces to

$$-\frac{\partial \Phi_{\text{eff}}}{\partial X} = 0, \qquad -\frac{\partial \Phi_{\text{eff}}}{\partial Y} = 0 \qquad (2.17)$$

at a Lagrange point, assuming a test mass at rest in the rotating reference frame. Using the effective potential from Equation 2.9 with $\Omega^2 = \mu/a^3$, the equation of motion at a Lagrange point reduces to

$$\frac{X}{a^3} - (1-f)\frac{X+fa}{r_1^3} - f\frac{X-(1-f)a}{r_2^3} = 0 \qquad (2.18)$$

in the X direction, and

$$\frac{Y}{a^3} - (1-f)\frac{Y}{r_1^3} - f\frac{Y}{r_2^3} = 0 \qquad (2.19)$$

---

[2] Three of the Lagrange points were first found by Euler; however, Joseph-Louis Lagrange bagged the complete set several years later, so he is credited in the name.

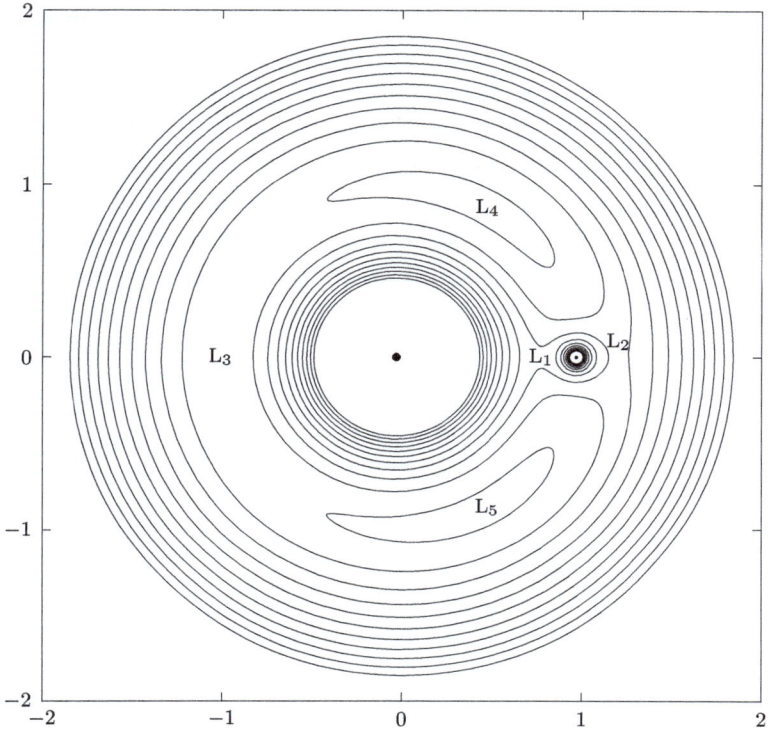

Figure 2.3 Equipotential contours of the effective potential in a restricted three-body system with $f = 0.03$. The units are chosen such that $\mu = 1$, $a = 1$, and thus $\Omega = 1$. The contours are drawn from $\Phi_{\text{eff}} = -1.5$ to $\Phi_{\text{eff}} = -2.25$, with spacing $\Delta\Phi_{\text{eff}} = 0.075$. The five Lagrange points are labeled.

in the Y direction. Equation 2.19 holds true when $Y = 0$; by symmetry, there is no acceleration in the Y direction for a test mass on the X axis. However, Equation 2.19 also holds true when $r_1 = r_2 = a$, as you can verify by substitution. That is, there is no acceleration in the Y direction for a test mass at one vertex of an equilateral triangle whose other vertices are defined by the massive primary and secondary. Moreover, when we substitute $r_1 = r_2 = a$ into Equation 2.18 (the equation of motion in the X direction), we find that

$$\frac{X}{a^3} - \frac{(1-f)X + f(1-f)a}{a^3} - \frac{fX - f(1-f)a}{a^3} = 0, \qquad (2.20)$$

as required for a Lagrange point. Thus, we have verified that the L_4 and L_5 points are true Lagrange points. In our rotating coordinate system, these non-collinear Lagrange points are at $X/a = 1/2 - f$ and $Y/a = \pm\sqrt{3}/2$, as shown in Figure 2.3.

The non-collinear Lagrange points L_4 and L_5 are always at a distance a from both the primary and the secondary, regardless of the secondary mass fraction f. However, the distances of the collinear Lagrange points from the primary and secondary do depend on f. We have seen that the acceleration in the Y direction is zero everywhere along the X axis; to locate the collinear Lagrange points, we must find the three points on the X axis where the acceleration in the X direction also vanishes. When the test mass is confined to the X axis, its distance from the primary is $r_1 = |X + fa|$ and its distance from the secondary is $r_2 = |X - (1 - f)a|$. Using these distances in the equation of motion in the X direction (Equation 2.18), we find that the collinear Lagrange points must satisfy the relation

$$\frac{X}{a} + \frac{1-f}{(X/a+f)^2} + \frac{f}{(X/a-1+f)^2} = 0 \qquad \left[\frac{X}{a} < -f\right], \tag{2.21}$$

$$\frac{X}{a} - \frac{1-f}{(X/a+f)^2} + \frac{f}{(X/a-1+f)^2} = 0 \qquad \left[-f < \frac{X}{a} < 1-f\right], \tag{2.22}$$

$$\frac{X}{a} - \frac{1-f}{(X/a+f)^2} - \frac{f}{(X/a-1+f)^2} = 0 \qquad \left[\frac{X}{a} > 1-f\right]. \tag{2.23}$$

Each of these equations can be written as a quintic equation, in which the highest-order term is $\propto X^5$. A numerical quintic solver can be used to solve the equations, showing that there is one solution for X in each of the three regions.

Equations 2.21 through 2.23 are not entirely impervious to analytic solution. As an example, consider Equation 2.22, whose solution gives the location of the L_1 point, between the primary and secondary. In the special case of an equal-mass binary, with $f = 1/2$, this equation can be written as

$$\frac{X}{a} - \frac{1/2}{(X/a+1/2)^2} + \frac{1/2}{(X/a-1/2)^2} = 0, \tag{2.24}$$

which has the solution $X/a = 0$. Thus, in an equal-mass binary, the L_1 point is located at the barycenter, exactly midway between the two massive objects. Like Buridan's ass exactly midway between two equally delicious haystacks, the test mass has no reason to move one way or another. An additional solution for the location of the L_1 point can be found in the limit $f \ll 1/2$, when the secondary is much less massive than the primary. In such a case, we expect that the equilibrium L_1 point will be closer to the low-mass secondary than to the higher-mass primary. Expressed in terms of the dimensionless distance of the test mass from the secondary, $s \equiv r_2/a = 1 - f - X/a$, Equation 2.22 becomes

$$-s + (1-f) - \frac{1-f}{(1-s)^2} + \frac{f}{s^2} = 0. \tag{2.25}$$

This equation can be rearranged to be slightly more illuminating:

$$f\left[\frac{1}{s^2} - s\right] = (1 - f)\left[\frac{1}{(1-s)^2} - (1-s)\right]. \tag{2.26}$$

Since $f \leq 1/2$ is the fraction of the mass in the secondary and $s = r_2/a$ is the dimensionless distance from the test mass to the secondary, this means that $1 - f \geq 1/2$ is the fraction of the mass in the primary and $1 - s = r_1/a$ is the dimensionless distance from the test mass to the primary.

Admittedly, Equation 2.26 is still a quintic equation. However, given our expectation that $s \ll 1$ when $f \ll 1/2$, we can make the approximation $1/s^2 \gg s$ on the left-hand side of Equation 2.26 and $(1-s)^{-2} \approx 1+2s$ on the right-hand side, leaving

$$\frac{f}{s^2} \approx (1 - f)3s, \tag{2.27}$$

or

$$s \equiv \frac{r_2}{a} \approx \left[\frac{f}{3(1-f)}\right]^{1/3}. \tag{2.28}$$

For the $f = 0.03$ binary shown in Figures 2.1 through 2.3, this places the L_1 point at a distance $r_2 \approx 0.22a$ from the secondary and $r_1 \approx 0.78a$ from the primary. You might (legitimately) argue that 0.22 is not extremely tiny compared to unity. If you want a more accurate approximation of the location of the L_1 point, expanding to higher-order terms gives the result

$$\frac{r_2}{a} \approx \alpha - \frac{1}{3}\alpha^2, \tag{2.29}$$

where

$$\alpha = \left[\frac{f}{3(1-f)}\right]^{1/3}. \tag{2.30}$$

For an $f = 0.03$ binary, this places the L_1 point at a distance $r_2 = 0.208a$ from the secondary and $r_1 = 0.792a$ from the primary.

A similar analysis for the L_2 point, starting from Equation 2.23, shows that it lies at a distance

$$\frac{r_2}{a} \approx \alpha + \frac{1}{3}\alpha^2 \tag{2.31}$$

from the secondary, on the far side from the L_1 point. Finally, the L_3 point lies on the far side of the primary, at a distance

$$\frac{r_1}{a} \approx 1 - \frac{7}{4}\alpha^3 \approx 1 - \frac{7}{12}f \tag{2.32}$$

from the primary. In the limit that $\alpha \ll 1$, the L_1 and L_2 points lie on a sphere of radius

$$r_{\mathrm{H}} = a \left[\frac{f}{3(1-f)} \right]^{1/3} = a \left(\frac{m_2}{3m_1} \right)^{1/3} \qquad (2.33)$$

centered on the secondary. This sphere is known as the **Hill sphere** of the secondary, named after the astronomer George William Hill. The Hill sphere can be thought of as the volume within which a moon of the secondary remains stably bound to the secondary, with only a minor perturbative effect from the primary. The concept of the Hill sphere can be extended to systems in which the orbit of the binary has a non-zero eccentricity e. In this case, the **Hill radius** can be generalized to

$$r_{\mathrm{H}} \approx a(1-e) \left(\frac{m_2}{3m_1} \right)^{1/3}. \qquad (2.34)$$

Relatively massive secondaries on large, nearly circular orbits around their primary have larger Hill spheres, and thus can potentially harbor larger families of satellites. The Hill radius of moon-rich Jupiter, for instance, is 290 times that of moon-deprived Mercury.

The Hill radius of the Earth as it orbits the Sun is

$$r_{\mathrm{H},\oplus} \approx 0.0098 \, \mathrm{au} \approx 1.47 \times 10^6 \, \mathrm{km}, \qquad (2.35)$$

about four times the semimajor axis of the Moon's orbit. Although the Moon is well inside the Earth's Hill radius, and will never spontaneously wander away to visit the Sun, the perturbing gravitational force of the Sun is sufficient to distort the Moon's orbit. The Sun's gravitational force causes eastward precession of the line of apsides of the Earth–Moon orbit, with a period of $\sim$9 yr. It also causes westward precession of the line of nodes (the intersection of the Earth–Moon orbital plane with the Sun–Earth orbital plane), with a period of $\sim$19 yr. The reason for the opposite directions and different periods is not intuitively obvious, even for a great physicist; according to Isaac Newton, "his head never ached but when he was studying the lunar irregularities."

2.3 Trojans, Tadpoles, and Horseshoes

Let us quietly tiptoe away from a problem that gave Newton a headache, and examine the slightly less brain-melting problem of the stability of the Lagrange points. The L_4 and L_5 points are maxima in the effective potential, as shown graphically in Figure 2.3. At the position of the L_4 and L_5 points, the effective potential (Equation 2.9) has the value

$$\Phi_{\mathrm{eff}}(L_4) = \Phi_{\mathrm{eff}}(L_5) = -\frac{3\mu}{2a} \left(1 - \frac{f}{3} + \frac{f^2}{3} \right), \qquad (2.36)$$

representing the global maxima in Φ_{eff}. The collinear Lagrange points, by comparison, are saddle points, representing maxima in Φ_{eff} as you move along the X axis, but minima as you move in the perpendicular direction. *If the acceleration on a test mass were due solely to the potential* Φ_{eff}, we would conclude that all the Lagrange points were unstable. An infinitesimal step (dX, dY) away from the maximum or the saddle point would result in an acceleration away from the Lagrange point. However, the situation is complicated by the presence of the Coriolis term in the equation of motion (Equations 2.7 and 2.8).

Empirical evidence suggests that the Coriolis term stabilizes the L_4 and L_5 Lagrange points. Near the L_4 and L_5 points of the Sun–Jupiter system, there exist large numbers of **trojan asteroids**, often called "trojans" for short. No celestial objects are found near the L_1, L_2, and L_3 points of the Sun–Jupiter system, hinting that these Lagrange points are unstable. Asteroids near Jupiter's L_4 and L_5 points are called "trojans" because the first such objects discovered were given names from the *Iliad*, Homer's epic poem about the Trojan War: 588 Achilles and 624 Hektor are near the L_4 point, while 617 Patroclus is near the L_5 point.[3]

A rigorous mathematical analysis of the Lagrange points is fairly straightforward – but is too long to fit into the margin of this book. The basic approach is to do a linear perturbation analysis, in which a test mass is moved an infinitesimal distance (dX, dY) from a Lagrange point. If the resulting linearized equation of motion (including the Coriolis term) predicts oscillatory motion, the Lagrange point is stable. If it predicts exponentially growing displacement, the Lagrange point is unstable. For the collinear Lagrange points, L_1, L_2, and L_3, linear perturbation theory predicts instability. For the L_3 point, on the far side of the primary, the e-folding time for the displacement of the test mass is $\tau \sim 0.3\mathcal{P}/f^{1/2}$, where $\mathcal{P}$ is the orbital period of the massive binary and f is the fraction of the total mass in the secondary. A low-mass "counter-Earth," at the L_3 point of the Sun–Earth system, would thus have $\tau \sim 200\,\text{yr}$. For the L_1 and L_2 points, the e-folding time is shorter and is independent of f, with $\tau \sim 0.06\mathcal{P}$. It is sometimes useful to park artificial satellites near the Sun–Earth L_1 and L_2 points. The L_1 point is useful for solar telescopes such as the *Solar and Heliospheric Observatory* (*SOHO*), while the L_2 point is useful for telescopes that you want to point away from the Sun, such as the *James Webb Space Telescope* (*JWST*). To keep a satellite near the Sun–Earth L_1 or L_2 points, a thruster on the satellite must fire at intervals of ~ 1 month to provide a non-gravitational restoring force.

---

[3] Initially, Jupiter's trojans were all named after characters from the *Iliad* and other poems about the Trojan War. However, now that $>10^4$ trojans are known, astronomers have exhausted the names from those poems and have switched to naming Jupiter's trojans after notable Olympic and Paralympic athletes.

Linear perturbation analysis of the L_4 and L_5 points leads to a more complicated result. It turns out that the L_4 and L_5 points are stable if the division of mass between the secondary and primary obeys the relation

$$f(1-f) \leq \frac{1}{27}. \qquad (2.37)$$

Solving this quadratic equation, and adopting the solution with $f < 1/2$, we find that for the L_4 and L_5 points to be stable, the fraction f of the total mass in the secondary must have an upper limit

$$f \leq \frac{1}{2} - \frac{\sqrt{69}}{18} \approx 0.0385. \qquad (2.38)$$

Within the solar system, most binary systems on nearly circular orbits have $f < 0.0385$. One exception is the Pluto–Charon system, with $f \approx 0.104$; don't expect to find any Pluto–Charon trojans. Every other planet–moon or Sun–planet pairing, however, has a sufficiently small value of f to support stable trojans.

Strictly speaking, the stability of the L_4 and L_5 points for $f < 0.0385$ applies only to an isolated three-body system. Within the solar system, the L_4 and L_5 points of planet–moon systems tend to be destabilized by the gravitational effect of the massive Sun. It is only when you get as far as Saturn, ~9.6 au from the Sun, that stable "trojan moons" are found. The binary consisting of Saturn and its moon Tethys ($f \approx 1.1 \times 10^{-6}$, $a \approx 2.0 \times 10^{-3}$ au) has a pair of trojan moons; Telesto is at the L_4 point and Calypso is at the L_5 point of the Saturn–Tethys system.

Throughout the inner solar system, the L_4 and L_5 points of the Sun–planet binaries tend to be destabilized in the long run by the perturbative effect of (relatively) nearby planets. It is the outer solar system that is the primary home of trojan asteroids. In addition to the swarms of trojans leading and following Jupiter, there are dozens of known Neptune trojans.[4]

Figure 2.4 shows the location, at a particular moment in time, of the trojan asteroids of Jupiter. Note that the trojans are not in a tight clump surrounding the L_4 and L_5 points, but are spread through an angle of ~60° as seen from the Sun's location. This is because the Coriolis terms don't stabilize test masses by slamming them straight back to the L_4 or L_5 point; instead, the Coriolis effect guides each test mass along a **tadpole orbit** around the relevant Lagrange point.

To visualize a tadpole orbit, imagine a trojan that is exactly at the L_4 potential maximum (Figure 2.3).[5] We displace it very slightly, by an amount $+dY$ in the positive Y direction. As the trojan starts to "roll down the potential hill,"

---

[4] The discovery of trojans at the Sun–Neptune L_5 point has been hindered by the annoying coincidence that it currently lies in the same direction as the center of our galaxy.
[5] Alternatively, you can place the trojan at L_5; the analysis is similar in both cases.

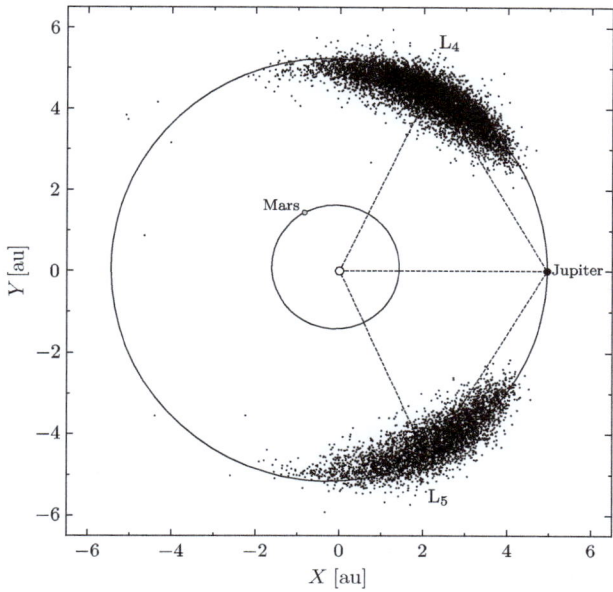

Figure 2.4 Locations of the known trojan asteroids of Jupiter. Jupiter (and the other planets) orbit in a counterclockwise direction in this plot.

it acquires a speed dY/dt in the positive Y direction. The Coriolis term in Equation 2.7 then creates an acceleration d^2X/dt^2 in the positive X direction. This causes the trojan to curve around in a clockwise sense, opposite to the counterclockwise orbital motion of the massive binary in Figure 2.3. In the limit of a minuscule initial displacement, the trojan's tadpole orbit is an ellipse centered on the L_4 (or L_5) point; when $f \ll 1$, the eccentricity of the counter-rotating tadpole orbit is $e_{\text{tad}} \approx 1 - 3f/2$, and its period is $\mathcal{P}_{\text{tad}} \sim 0.4\mathcal{P}/f^{1/2}$. For Jupiter trojans, this means the aspect ratio of a small tadpole orbit (minor axis divided by major axis) is $(b/a)_{\text{tad}} \sim (3f)^{1/2} \sim 0.054$, while its period is $\mathcal{P}_{\text{tad}} \sim 13\mathcal{P}_{\text{jup}} \sim 150\,\text{yr}$.

Given the shape of the effective potential, the counter-rotating motion about the L_4 and L_5 points develops an elongated tail when the displacement from the potential maximum is not infinitesimally tiny; this long tail is what prompted the name "tadpole" for this type of orbit. Figure 2.5 shows an early numerical calculation of tadpole orbits around the L_5 point of the Sun–Jupiter system. Notice that larger initial displacement leads to a tadpole that curves along a circular arc. For a sufficiently large initial displacement, the tail of a tadpole orbit can stretch as far as the L_3 point. When the length of the tadpole orbit stretches to its maximum value of $\sim 2a$, its width becomes $\sim 4f^{1/2}a$. For Jupiter trojans, this gives a maximum tadpole width of $\sim 0.1a_{\text{jup}} \sim 0.5\,\text{au}$. (Compare this to the actual width of the cloud of trojan asteroids in Figure 2.4.)

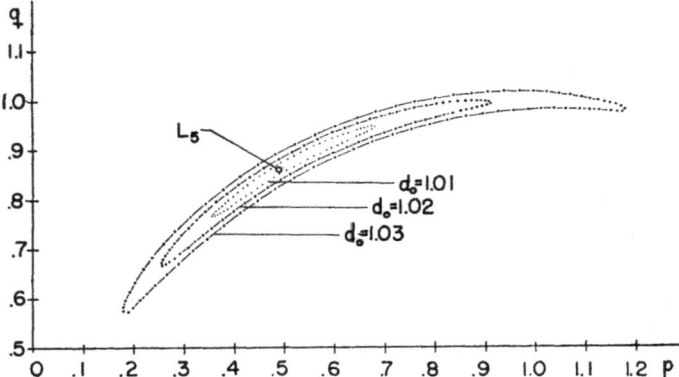

Figure 2.5 Tadpole orbits around the L_5 point, as computed by Eugene Rabe in 1961 using "an IBM 650 electronic computer." Rabe uses the convention that $a = 1$, that the primary is at $(p, q) = (1, 0)$, and that the secondary is on the left. The parameter d_0 is a measure of the initial distance between the test mass and primary. [Rabe 1961, Fig. 1]

What happens if we elongate a tadpole's tail until it stretches *past* the L_3 point? In that case, the orbit becomes a **horseshoe orbit**, enclosing both the L_4 and L_5 points, in addition to the L_3 point. Figure 2.6 shows a calculated horseshoe orbit for the Sun–Jupiter system.[6] It is easier to understand horseshoe orbits in the limit $f \ll 1/2$, and the secondary mass is tiny compared to that of the primary. In this case, when the test mass is at the L_3 point, it is twice as far from the secondary as from the primary. The gravitational force of the primary on the test mass is thus greater than that of the secondary by a factor $\sim 4/f$. At the L_3 point of the Sun–Earth system, for instance, the Earth provides a perturbation of only one part per million to the Sun's gravitational acceleration.

To visualize a horseshoe orbit in the case $f \ll 1/2$, imagine a test mass that is exactly at the L_3 point (Figure 2.3), at rest in the rotating frame of reference and thus having an orbital speed $\Omega = \mu^{1/2}/a^{3/2}$ in the inertial frame. Now we displace the test mass by a distance dX in the direction opposite to the primary, while tweaking its velocity slightly to ensure that it is on a circular orbit in the inertial frame. In our $f \ll 1/2$ approximation, the angular speed of the test mass in the inertial frame is now

$$\Omega \approx \frac{\mu^{1/2}}{(a + dX)^{3/2}} \approx \frac{\mu^{1/2}}{a^{3/2}} \left(1 - \frac{3}{2}\frac{dX}{a}\right). \tag{2.39}$$

---

[6] The term "horse-shoe orbit" was first used by Ernest Brown in 1911. During that pre-WWI era, the United States reached its maximum horse population, with $\sim$20 million horses (as compared to $\sim$90 million humans in the USA at that time).

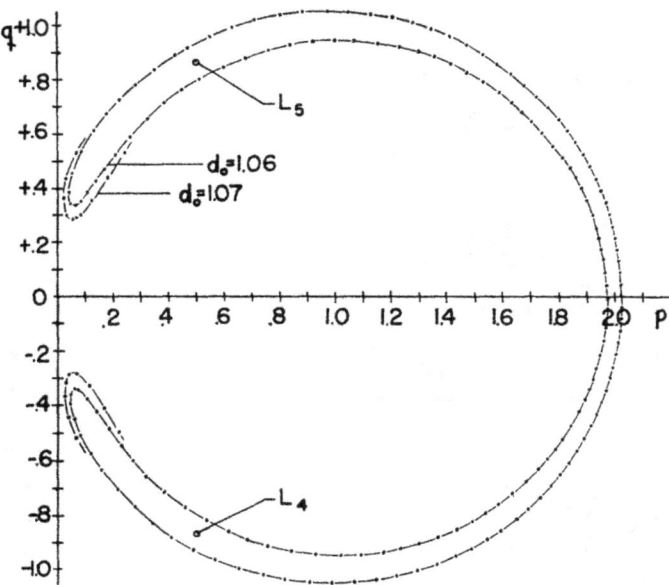

Figure 2.6 Horseshoe orbits around the L_4 and L_5 points, as computed by Eugene Rabe in 1961. The units and coordinates are the same as in Figure 2.5. [Rabe 1961, Fig. 4]

Thus, the angular speed of the test mass is slightly smaller than $\mu^{1/2}/a^{3/2}$. In the rotating frame, the test mass therefore drifts gradually past the leading L_4 point toward the secondary. However, as the test mass approaches the Hill sphere of the secondary, the gravitational force of the secondary starts to have a significant influence on the test mass. In the inertial frame, the secondary is approaching the test mass from behind; thus, the gravitational attraction of the secondary *decreases* the orbital angular momentum of the test mass, moving it onto a smaller orbit around the primary, with $r_1 < a$. Now the angular speed of the test mass is slightly greater than $\mu^{1/2}/a^{3/2}$, and it drifts gradually back past the L_4 point and the L_3 point toward the trailing L_5 point. Eventually the gravity of the secondary starts to influence the test mass again, this time *increasing* the orbital angular momentum of the test mass, and causing it to move back onto a larger orbit. The motion of the test mass along the resulting horseshoe orbit continues indefinitely.

The calculated tadpole and horseshoe orbits in Figures 2.5 and 2.6 represent the special cases when the orbits are simple closed loops. Real tadpole and horseshoe orbits are generally not as simple. The Earth, for instance, has a number of co-orbital bodies on horseshoe orbits; the largest such object, 3753 Cruithne, is $\sim$5 km in diameter, but most are much smaller. Thus, the horseshoe co-orbital bodies are excellent approximations to a test mass.

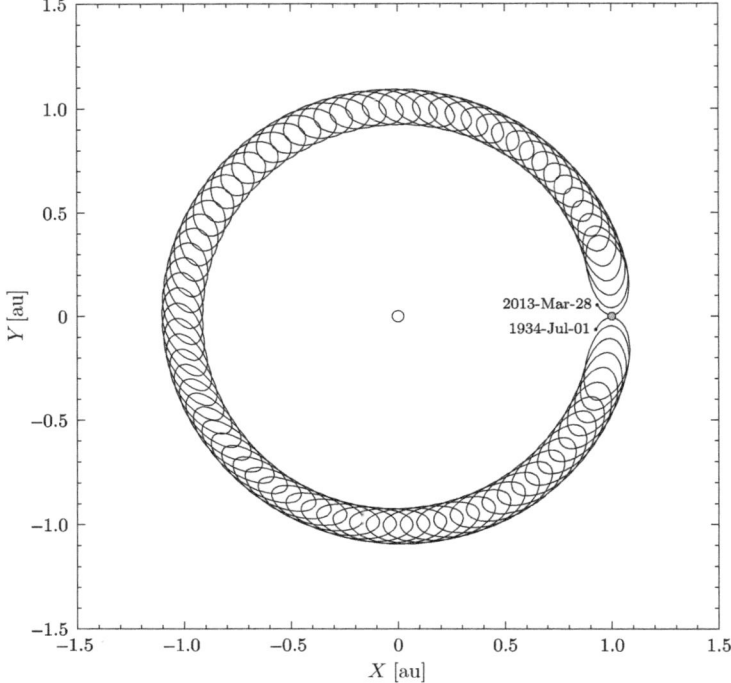

Figure 2.7 Horseshoe orbit of the object 2013 BS$_{45}$, integrated from 1934, just before it made a close approach to the Earth, to 2013, when it was discovered during its next close approach. [Calculated using the JPL Horizons system]

However, the Sun–Earth orbit is not perfectly circular ($e = 0.017$), and the co-orbital bodies are not exactly in the Sun–Earth orbital plane. Figure 2.7 shows the horseshoe orbit of the co-orbital object 2013 BS$_{45}$, which was discovered in 2013 on a close approach to the Earth; its inclination is only $i \approx 0.8°$, so it is nearly coplanar with the Earth's orbit. In the rotating frame, 2013 BS$_{45}$ oscillates in the radial direction with an amplitude $\Delta R \sim 0.16$ au and a period $\mathcal{P} \sim 1$ yr. This reflects the fact that in the inertial frame, over the course of a single year, the orbit of 2013 BS$_{45}$ around the Sun is well fitted by an ellipse with semimajor axis $a \sim 1$ au and eccentricity $e \sim 0.08$. In the rotating frame, the time for 2013 BS$_{45}$ to drift from one end of the horseshoe to the opposite end and back again is $\mathcal{P}_{\mathrm{horse}} \sim 160$ yr.

2.4 Hierarchical Triples

The restricted three-body problem has many applications in our solar system. However, our galaxy is full of three-body systems which do not even remotely obey the limitations of the restricted three-body problem. Of the 100 closest

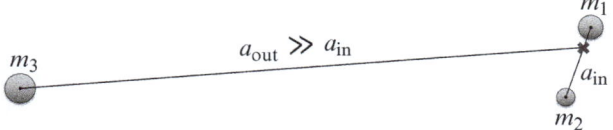

Figure 2.8 A hierarchical triple system. The primary (mass m_1) and secondary (mass $m_2 < m_1$) constitute the tight inner binary. The tertiary (mass m_3) is at a distance $a_{\text{out}} \gg a_{\text{in}}$ from the inner binary.

stellar systems to the Sun, seven are known to be triple star systems; thus, triple stars are not rare. However, if we look at the nearby α Centauri system, we find that the primary and secondary, α Cen A and α Cen B, are on an eccentric orbit ($e = 0.52$) rather than a circular orbit. The tertiary, Proxima Centauri, is on an orbit inclined by $i \sim 30°$ relative to that of the A + B binary, rather than orbiting in the same plane. In addition, Proxima Centauri has a mass $\sim 12\%$ that of α Cen B, rather than being negligible in mass.

Since the results found from the restricted three-body problem don't necessarily apply to triple star systems, we need to develop a new set of tools. One empirical result found from studying triple star systems is that they are all **hierarchical triples**; that is, they contain one tightly bound binary with semimajor axis a_{in} plus a third star at a distance $a_{\text{out}} \gg a_{\text{in}}$ from the barycenter of the tight inner binary. An example of such a hierarchical triple is shown in Figure 2.8. (For clarity, Figure 2.8 has a relatively modest ratio of a_{out} to a_{in}. The α Centauri system, to take a real-universe example, has $a_{\text{in}} = 23.3$ au and $a_{\text{out}} \approx 8700$ au $\approx 370 a_{\text{in}}$.)

Within a hierarchical triple, the more massive member of the inner binary, with mass m_1, is called the primary; the less massive member of the inner binary ($m_2 < m_1$) is called the secondary. The outer third star, at $a_{\text{out}} \gg a_{\text{in}}$, is called the tertiary, regardless of its mass. In studying hierarchical triples, we toss away the restriction that the tertiary must lie in the orbital plane of the inner binary. We jettison the restriction that the inner binary must have a circular orbit. We discard the restriction that the mass of the tertiary must be negligibly small. We are, however, adding the restriction that $a_{\text{out}} \gg a_{\text{in}}$. More precisely, when the eccentricity e_{out} of the tertiary's orbit is non-zero, long-term stability requires that the periapsis distance $r_{\text{pe}} = a_{\text{out}}(1 - e_{\text{out}})$ between the tertiary and the inner binary be much larger than the typical distance a_{in} between the primary and secondary.

In the limit

$$\frac{a_{\text{in}}}{a_{\text{out}}} \ll 1 - e_{\text{out}}, \tag{2.40}$$

the gravitational potential experienced by the tertiary is nearly equal to the Keplerian potential of a spherical object with mass $m_1 + m_2$ located at the barycenter of the inner binary. Thus, the orbit of the tertiary is very nearly a perfect Keplerian ellipse. However, if the mass of the tertiary is sufficiently large, it can have a significant perturbative effect on the inner binary.

One way in which a distant tertiary can affect an inner binary system is through the **von Zeipel–Lidov–Kozai (ZLK) mechanism**, in which the outer tertiary causes periodic variations in the inclination and eccentricity of the inner binary's orbit. In older works, this effect is called the Lidov–Kozai mechanism, the Kozai–Lidov mechanism, or simply the Kozai mechanism. Although the ZLK mechanism was first described by Hugo von Zeipel in 1910, his paper languished uncited for nearly a century. In 1961, Mikhail Lidov independently discovered the ZLK mechanism while analyzing how satellites in low Earth orbit are perturbed by the Moon. Yoshihide Kozai, after hearing a presentation by Lidov, developed the idea in the context of asteroids perturbed by Jupiter, publishing his results in 1962.

The simplest possible analysis of the ZLK mechanism starts with the assumption that the secondary is of negligible mass compared to the primary and to the distant perturbing tertiary. This is an excellent approximation in the (Sun + comet) + Jupiter system studied by von Zeipel, the (Earth + satellite) + Moon system studied by Lidov, and the (Sun + asteroid) + Jupiter system studied by Kozai. In the limit that $m_2 \to 0$, we can assume that the barycenter of the inner binary is at a location indistinguishable from the center of the primary. The secondary is on an orbit of eccentricity e about the primary; this orbit is inclined by an angle i relative to the fixed orbital plane of the perturbing tertiary; the eccentricity e and inclination i of the secondary's orbit are the properties altered by the ZLK mechanism.

In an inertial frame of reference with arbitrary origin $(0, 0, 0)$, the locations of the primary, the test-mass secondary, and the perturbing tertiary are $\vec{x}_1, \vec{x}_2$, and $\vec{x}_3$. The position of the secondary relative to the primary is $\vec{r}_2 \equiv \vec{x}_2 - \vec{x}_1$; the position of the tertiary relative to the primary is $\vec{r}_3 \equiv \vec{x}_3 - \vec{x}_1$, as shown in Figure 2.9. The equation of motion for the primary, since m_2 is negligible, takes into account only the gravitational pull of the tertiary:

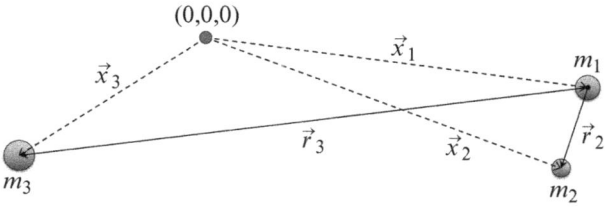

Figure 2.9 Relative positions of the primary, test-mass secondary, and tertiary in a hierarchical triple system.

$$\frac{d^2\vec{x}_1}{dt^2} = \mu_3 \frac{\vec{r}_3}{r_3^3}, \tag{2.41}$$

where $\mu_3 = Gm_3$. The equation of motion for the secondary, however, must take into account both the primary and the perturbing tertiary:

$$\frac{d^2\vec{x}_2}{dt^2} = -\mu_1 \frac{\vec{r}_2}{r_2^3} + \mu_3 \frac{\vec{r}_3 - \vec{r}_2}{|\vec{r}_3 - \vec{r}_2|^3}. \tag{2.42}$$

By subtracting Equation 2.41 from Equation 2.42, we find the equation of relative motion for the test-mass secondary relative to the primary:

$$\frac{d^2\vec{r}_2}{dt^2} = -\mu_1 \frac{\vec{r}_2}{r_2^3} + \mu_3 \left(\frac{\vec{r}_3 - \vec{r}_2}{|\vec{r}_3 - \vec{r}_2|^3} - \frac{\vec{r}_3}{r_3^3} \right). \tag{2.43}$$

This equation of relative motion can be written as

$$\frac{d^2\vec{r}_2}{dt^2} = -\vec{\nabla}(\Phi_K + \mathcal{R}), \tag{2.44}$$

where the gradient is with respect to the coordinate $\vec{r}_2$. The function Φ_K has the familiar Keplerian form

$$\Phi_K(r_2) = -\frac{\mu_1}{r_2}, \tag{2.45}$$

representing the gravitational potential of the primary. The function $\mathcal{R}$ is called the **disturbing function**, and represents the contribution from the perturbing tertiary:

$$\mathcal{R}(r_2) = -\frac{\mu_3}{|\vec{r}_2 - \vec{r}_3|} + \mu_3 \frac{\vec{r}_3 \cdot \vec{r}_2}{r_3^3} = -\frac{\mu_3}{|\vec{r}_3 - \vec{r}_2|} + \mu_3 \frac{r_2}{r_3^2} \cos\psi, \tag{2.46}$$

where ψ is the angle between $\vec{r}_2$ and $\vec{r}_3$. In its full form, the disturbing function is not very enlightening. However, in a hierarchical triple we can assume $r_2 \ll r_3$, even when the tertiary is at its closest approach to the primary. Thus, we follow in the footsteps of Adrien-Marie Legendre and do an expansion of $1/|\vec{r}_3 - \vec{r}_2|$ in terms of Legendre polynomials:

$$\frac{1}{|\vec{r}_3 - \vec{r}_2|} = \frac{1}{r_3} \sum_{\ell=0}^{\infty} \left(\frac{r_2}{r_3} \right)^{\ell} P_{\ell}(\cos\psi), \tag{2.47}$$

where ψ, as above, is the angle between $\vec{r}_2$ and $\vec{r}_3$. The first few Legendre polynomials are $P_0(x) = 1$, $P_1(x) = x$, and $P_2(x) = (3x^2 - 1)/2$. When we use this expansion in Equation 2.46, the $\ell = 1$ term cancels out, and the disturbing function can be written approximately as

$$\mathcal{R} \approx -\frac{\mu_3}{r_3} - \mu_3 \frac{r_2^2}{r_3^3} P_2(\cos\psi). \tag{2.48}$$

From a physics viewpoint, we are interested only in the gradient of the disturbing function with respect to $\vec{r}_2$; this gives the acceleration of the test mass due to the perturbing tertiary. Since the first term in Equation 2.48 is independent of $\vec{r}_2$, the disturbing function is usually renormalized as

$$\mathcal{R} \approx -\mu_3 \frac{r_2^2}{r_3^3} P_2(\cos \psi). \tag{2.49}$$

Since the leading term in this approximation is the $\ell = 2$ (quadrupole) term, Equation 2.49 is called the **quadrupole approximation** for the disturbing function.[7] In the quadrupole approximation, the acceleration of the test mass by the perturbing tertiary is smaller than its acceleration by the primary; the approximate ratio of the accelerations is

$$\frac{|\vec{\nabla}\mathcal{R}|}{|\vec{\nabla}\Phi_K|} \sim \frac{\mu_3}{\mu_1} \left(\frac{r_2}{r_3}\right)^3 \sim \frac{\mu_3}{\mu_1} \left(\frac{a_{\text{in}}}{a_{\text{out}}}\right)^3. \tag{2.50}$$

In a hierarchical triple, the effect of the perturbing tertiary is small; this means that for a single orbit of the test-mass secondary, the Keplerian approximation is valid. That is, over a single orbit, we can treat the semimajor axis a_{in} and eccentricity e_{in} of the secondary's orbit as being nearly constant, as well as its inclination i relative to the orbital plane of the distant tertiary.

However, we also need to take the longer view, and consider what happens on timescales much longer than the secondary's orbital period,

$$\mathcal{P}_{\text{in}} = \frac{2\pi}{\mu_1^{1/2}} a_{\text{in}}^{3/2}. \tag{2.51}$$

In studying orbital motion, conservation laws are often helpful. If the perturbing tertiary were whisked away to infinity, then the unperturbed orbit of the test-mass secondary would conserve both specific angular momentum, $j = |\vec{r}_1 \times \vec{v}_1|$, and specific energy, $\epsilon = v_1^2/2 - \mu_1/r_1$. For closed elliptical orbits, these conserved quantities are (Equations 1.48 and 1.50)

$$j = \mu_1^{1/2} a_{\text{in}}^{1/2} (1 - e_{\text{in}}^2)^{1/2}, \tag{2.52}$$

$$\epsilon = -\frac{\mu_1}{2a_{\text{in}}}. \tag{2.53}$$

We are fixing the masses of all three bodies in our system, so we can assume μ_1 is constant. If the perturbing tertiary is on an orbit of finite (but large) size, its gravitational attraction is not effective at doing net work on the test mass. Thus, even on timescales long compared to $\mathcal{P}_{\text{out}}$, we can assume the specific energy ϵ of the test mass is constant. This implies, from Equation 2.53, that the average distance a_{in} of the test mass from the primary is constant.

---

[7] Warning: The quadrupole approximation breaks down when the perturbing tertiary is not very distant and when the orbits of the secondary and tertiary are both highly eccentric.

However, since the orbit of the perturbing tertiary is inclined at an angle i relative to that of the secondary, it can change the angular momentum j of the inner binary. Only the component of $\vec{j}$ perpendicular to the orbital plane of the tertiary is conserved. That is, the conserved quantity is (using Equation 2.52)

$$j_z = \mu_1^{1/2} a_{\text{in}}^{1/2} (1 - e_{\text{in}}^2)^{1/2} \cos i. \qquad (2.54)$$

Since the gravitational parameter μ_1 and orbital semimajor axis a_{in} are constant in this problem, a conserved quantity for the orbit of the test-mass secondary is the **Kozai integral**,

$$f_z \equiv (1 - e_{\text{in}}^2)^{1/2} \cos i. \qquad (2.55)$$

No matter how the perturbing tertiary alters the secondary's orbit, the Kozai integral f_z must remain constant.

Suppose the orbit of the test-mass secondary starts out nearly circular ($e_{\text{in}} \approx 0$); in this case, increasing eccentricity is coupled with an increase in $\cos i$, and thus a decrease in the inclination i of the secondary's orbit relative to the tertiary's orbit. Such a coupling between increasing eccentricity and decreasing inclination is seen in Figure 2.10, which shows how the Earth's orbit would be altered if the Sun had a companion star at a

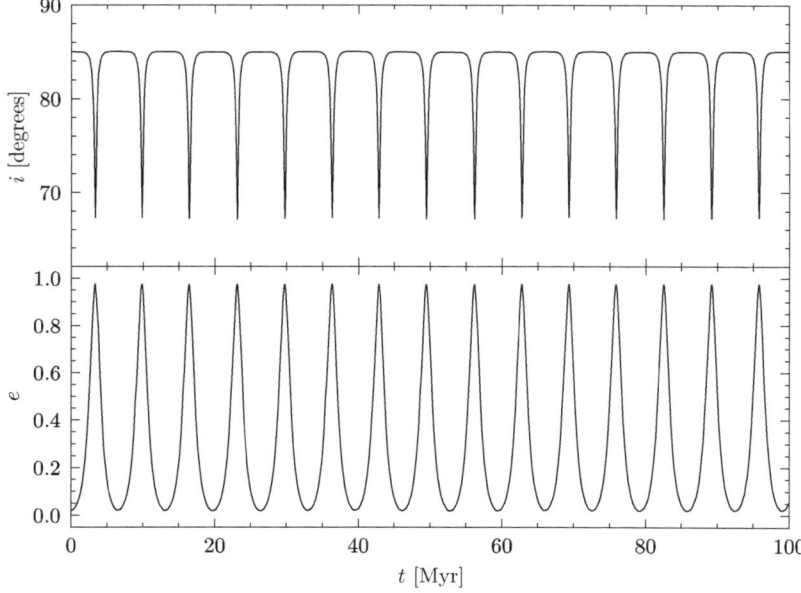

Figure 2.10 Computed ZLK oscillations for the Earth ($a_{\text{in}} = 1$ au) given a perturbing tertiary in the outer solar system with $m_3 = m_1 = 1\,\text{M}_\odot$, $a_{\text{out}} = 200$ au, and $e_{\text{out}} = 0.1$. The initial inclination of the Earth's orbit relative to the tertiary's is $i = 85°$.

distance $a_{out} = 200$ au. The **ZLK oscillations** shown in Figure 2.10, trading off inclination and eccentricity, can occur only when

$$f_z^2 = (1 - e_{in}^2) \cos^2 i < \frac{3}{5}. \tag{2.56}$$

This means that if the orbit of the secondary is originally circular, it will remain circular if its inclination i is initially less than

$$i_{crit} = \cos^{-1}(\sqrt{3/5}) \approx 39.2°. \tag{2.57}$$

Thus, changing the eccentricity and inclination of the secondary's orbit requires a distant perturbing tertiary at a high relative inclination.

If the secondary starts out on an orbit with a high inclination relative to the tertiary's orbit ($i \approx 90°$ and thus $\cos i \ll 1$), then the ZLK oscillations can drive the eccentricity to very high values. For instance, Figure 2.10 shows what would happen if the Earth, with initial eccentricity $e_{in} = 0.02$, were perturbed by a tertiary with initial inclination $i = 85°$, yielding a Kozai integral $f_z = 0.087$. For the initial conditions adopted in Figure 2.10, the inclination of the Earth's orbit relative to the tertiary's orbit drops as low as $i \approx 67°$. This drives the eccentricity of the Earth's orbit up to $e_{in} \approx 0.974$, leading to a perihelion distance $r_{pe} \approx 0.026$ au $\approx 5.6 R_\odot$, which would make the Earth uncomfortably warm once per year. (By adopting a different eccentricity and argument of periapsis for the perturbing tertiary, the inclination can be driven as low as $i \sim 0°$, which would lead to $e_{in} \approx 0.9962$ and $r_{pe} \approx 0.0038$ au $\approx 0.8 R_\odot$. Goodbye, Earth.)

Although ZLK oscillations can have profound influences on a planet, and any life on that planet, we still need to investigate the timescale on which the oscillations happen. In Figure 2.10, the characteristic timescale t_{ZLK} is the time that elapses between minima in the eccentricity of the secondary's orbit. For the parameters chosen, the minima occur quite regularly, on a timescale of several million times the orbital period of the secondary. The basic timescale for ZLK oscillations, in terms of the orbital period $\mathcal{P}_{in}$ of the test-mass secondary, is

$$\frac{t_{ZLK}}{\mathcal{P}_{in}} \approx \frac{m_1 + m_3}{m_3} \left(\frac{\mathcal{P}_{out}}{\mathcal{P}_{in}}\right)^2 (1 - e_{out}^2)^{3/2}. \tag{2.58}$$

Using Kepler's third law (Equation 1.58), this can also be written as

$$\frac{t_{ZLK}}{\mathcal{P}_{in}} \approx \frac{m_1}{m_3} \left(\frac{a_{out}}{a_{in}}\right)^3 (1 - e_{out}^2)^{3/2}. \tag{2.59}$$

The calculation that went into Figure 2.10 assumed $m_3 = m_1$, $a_{out} = 200 a_{in}$, and $e_{out} = 0.1$. Thus, Equation 2.59 yields $t_{ZLK} \approx 8 \times 10^6 \mathcal{P}_{in}$, or about 8 Myr, in agreement with the numerical calculations.

Our basic assumption for stable hierarchical triples was that $a_{out}(1-e_{out}) \gg a_{in}$. Given this restriction, Equation 2.59 tells us that the only situation in which t_{ZLK} is not very much longer than the secondary's orbital period $\mathcal{P}_{in}$ is the case in which the perturbing tertiary is much more massive than the primary ($m_3 \gg m_1$). Such a case is found among the moons of Jupiter, where the perturbing tertiary is the Sun, with $m_3 = 1047m_1$. Consider, for instance, Jupiter's moon Carpo (named after one of the numerous daughters of the god Jupiter). The semimajor axis of Carpo's orbit is $a_{in} = 0.114$ au, about 32% of Jupiter's Hill radius. Carpo's orbit is highly inclined relative to the Sun–Jupiter orbit, with an average inclination $i \approx 53°$. Combined with the mean eccentricity $e_{in} \approx 0.425$ of Carpo's orbit, this gives a Kozai integral $f_z \approx 0.54$; since this is below the critical value $f_{crit} = (3/5)^{1/2} \approx 0.77$, Carpo undergoes ZLK oscillations. Given that the Sun–Jupiter orbit has $a_{out} = 5.20$ au and $e_{out} = 0.049$, Equation 2.59 tells us that

$$\frac{t_{ZLK}}{\mathcal{P}_{in}} \approx \frac{1}{1047} \left(\frac{5.20}{0.114} \right)^3 (0.9976)^{3/2} \approx 90. \tag{2.60}$$

Since the orbital period of Carpo is $\mathcal{P}_{in} = 1.25$ yr, we can see the orbit of Carpo undergo a complete ZLK cycle over the course of roughly a century. The maximum eccentricity reached by Carpo, calculated to be $e_{max} = 0.69$, places its periapsis at a distance $r_{pe} \approx 74\,R_{jup}$ from the center of Jupiter. Although this seems like a safe distance, it is only ~2.8 times the semimajor axis of Callisto, the outermost Galilean moon of Jupiter. Just as the Galilean moons significantly perturb each others' orbits (Section 3.5), they also have a perturbative effect on the orbit of Carpo.

Exercises

2.1 A planet with mass m_p orbits the Sun. Its orbital period is $\mathcal{P}_{orb,p}$, while its rotation period is $\mathcal{P}_{rot,p}$. You want to place an artificial satellite in a "stationary" orbit around this planet; that is, you want the satellite to be on a circular orbit in the equatorial plane of the planet, with an orbital period $\mathcal{P}_{orb,s}$ that is equal to the rotation period $\mathcal{P}_{rot,p}$ of the planet.
   (a) For the satellite's orbit to be stable, it must be smaller than the Hill radius r_H of the planet. For a given planetary mass m_p and orbital period $\mathcal{P}_{orb,p}$, for what range of rotation periods $\mathcal{P}_{rot,p}$ is a stable stationary orbit possible? [Assume a satellite of negligible mass.]
   (b) For which of the eight major planets of the solar system are stable stationary orbits impossible?
2.2 Consider the three-body problem in the limiting case $m_1 = m_2$ (that is, the degenerate case in which the primary and secondary are of equal mass).

(a) Demonstrate that if a test mass is displaced by a small amount δX from the L_4 point, with $V = 0$ in the corotating frame, its distance from the L_4 will initially increase exponentially with time.

(b) Compute the e-folding time τ for the increasing distance from the L_4 point. What is the value of τ, in years, for a pair of solar-mass stars separated by one astronomical unit?

3

Resonances and Chaos

One meteorologist remarked that if the theory were correct,
one flap of a sea gull's wings would be enough to alter the course
of the weather forever. The controversy has not yet been settled,
but the most recent evidence seems to favor the sea gulls.

Edward Lorenz (1917–2008)
"The Predictability of Hydrodynamic Flow" [1963]

A gravitationally bound two-body system (if the two bodies are spheres of constant mass) shows simple periodic motion. We have seen that a three-body system, even if we install restrictions for computational simplicity, can show a rich variety of behaviors. Tadpole orbits, horseshoe orbits, and ZLK oscillations are just a sampling of what can happen.

The Minor Planet Center of the International Astronomical Union has tallied over 1 200 000 minor planets in the solar system. With such a large number of test masses bound to the Sun and perturbed by the eight major planets, we have observed evidence for a rich diversity of dynamical behavior, including **resonances** and **chaotic orbits**. The increasing number of exoplanets with well-determined orbits gives us another toy box full of dynamical systems to play with.

3.1 Mean Motion Resonances

Within the solar system, many examples are found of objects whose orbits display commensurability. If two objects orbit the same primary, then their orbits are **commensurable** if the ratio of their orbital periods can be expressed as the ratio of two small, mutually prime integers. For historical reasons, the ratios are usually expressed in terms of the mean motion $n \equiv 2\pi/\mathcal{P}$ of the orbits, rather than their orbital period $\mathcal{P}$. The mean motion is sometimes expressed in units of degrees per unit time. However, we will stick with radians per unit time, in part because Kepler's third law (Equation 1.58) then has the simple form

$$a^3 n^2 = \mu, \tag{3.1}$$

where μ is the standard gravitational parameter of the system.

One famous example of commensurability is given by Neptune, with orbital period $\mathcal{P}_N = 164.79\,\text{yr}$, and Pluto, with $\mathcal{P}_P = 247.92\,\text{yr}$. This translates to mean motions of $n_N = 0.038\,13\,\text{yr}^{-1}$ and $n_P = 0.025\,34\,\text{yr}^{-1}$. The ratio of their mean motions is thus $n_N/n_P = 1.504$, which is close to, but not identical with, the ratio $3:2$. The trojans of Jupiter are commensurate with Jupiter's mean motion at a ratio of $1:1$. The small trojan moons Telesto and Calypso are similarly commensurate with Saturn's larger moon Tethys.

By definition, commensurability involves the ratios of small integers; in practice, "small" usually means "less than 20." This is not because astronomers are incapable of counting higher than the number of their fingers plus toes. Instead, it is because ratios involving small integers (such as $1:1$ or $3:2$) are linked to interesting resonant phenomena. If we state that the mean motions of Neptune and Pluto have the ratio $24\,792 : 16\,479$, that is simply a statement of how accurately we can measure the orbital periods of the two objects. However, if we state that their mean motions have the ratio $3:2$, that means any Sun–Neptune–Pluto configuration will recur after a time equal to $3\mathcal{P}_N = 2\mathcal{P}_P \approx 495\,\text{yr}$.

When two objects orbiting the same primary have commensurable orbital periods (or equivalently, commensurable mean motions), they are said to have a **mean motion resonance**. When one orbiting object is much more massive than the other, then the massive object is not significantly changed by the resonance; however, the orbit of the less massive object can be perceptibly modified. In the case of Neptune and Pluto, the mass of Neptune is 7800 times that of the dwarf planet Pluto. Thus, Neptune is barely perturbed by Pluto's gravity, but Pluto's orbit is affected by Neptune's gravity.

Because mean motion resonances occur throughout the solar system and beyond, there is an entire vocabulary devoted to describing them. When the less massive object is on the larger orbit, this is called an **exterior resonance**; for instance, Pluto is in a $3:2$ exterior resonance with Neptune. Many other objects in the Kuiper belt, known collectively as the "plutinos," are also in a $3:2$ exterior resonance with Neptune. There are also Kuiper belt objects in a $2:1$ exterior resonance with Neptune; these are known as "twotinos" (pronounced *too-teenos*, for the sake of the rhyme with "plutinos"). When the less massive object in a mean motion resonance is on the smaller orbit, this is called an **interior resonance**. Asteroids in the main asteroid belt between Mars and Jupiter can be in an interior resonance with Jupiter.

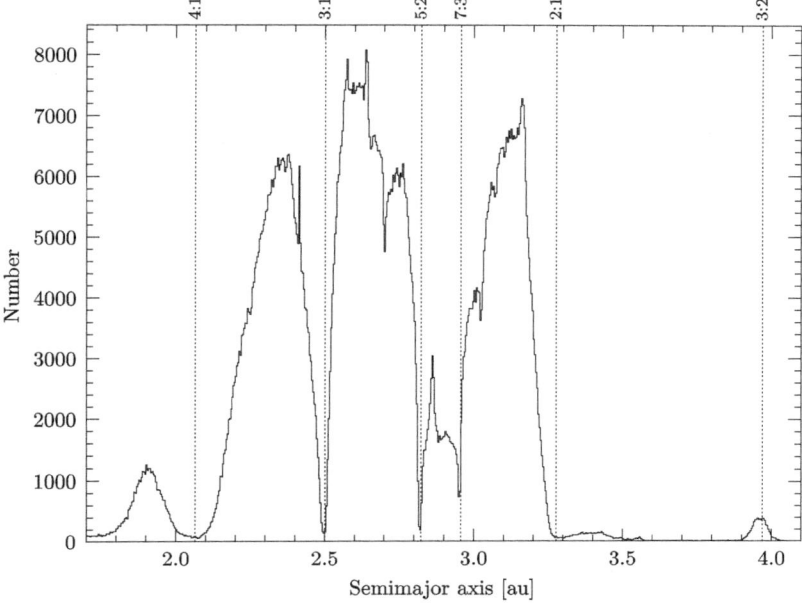

Figure 3.1 Number of known asteroids as a function of orbital semimajor axis a; the histogram bin width is $\Delta a = 0.004$ au. Vertical dotted lines correspond to the labeled interior resonances with Jupiter.

The asteroid belt, given the large number of asteroids whose orbits are known, is a rich source of information about interior resonances. A plot of the number of asteroids as a function of orbital semimajor axis (Figure 3.1) shows significant features where an asteroid has a mean motion resonance with Jupiter. Curiously, some of the resonances in the asteroid belt (such as the 4 : 1 and 3 : 1 resonances) are associated with a shortage of asteroids, while others (most notably the 3 : 2 resonance) are associated with an excess of asteroids. The resonances where there is a dip in asteroid number are known as the **Kirkwood gaps**. In the year 1866, when there were fewer than 100 asteroids with known orbits, Daniel Kirkwood identified gaps at semimajor axes corresponding to the 3 : 1 and 5 : 2 mean motion resonances. Other gaps, shown in Figure 3.1, were discovered later. The excess asteroids near the 3 : 2 mean motion resonance at $a \approx 3.97$ au are known as the **hildas**, after the asteroid 153 Hilda, the first object discovered at that resonance.[1]

To give a taste of the complications associated with resonances, consider a simplified version of the solar system in which we discard all planets other than Jupiter and Mars. We approximate Jupiter as being on a circular orbit

---

[1] 153 Hilda is a fairly large asteroid, with mean diameter $d \approx 170$ km. However, because it is a dark carbonaceous asteroid at a relatively large distance, it wasn't discovered until 1875, after smaller but flashier asteroids such as 5 Astraea and 17 Thetis.

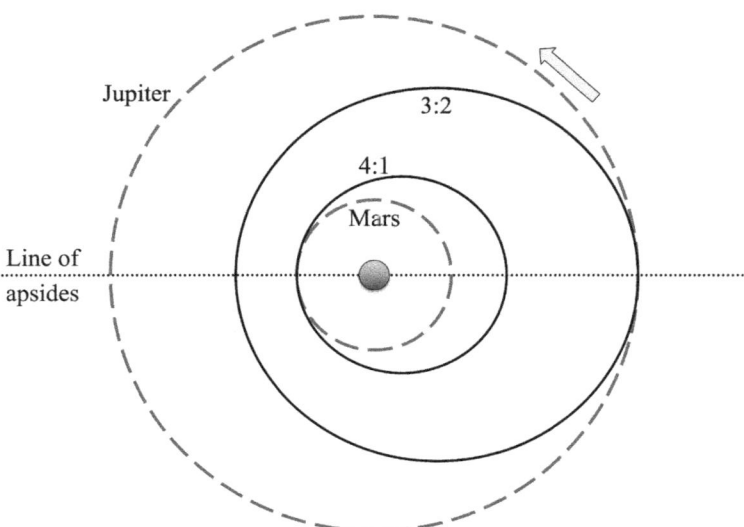

Figure 3.2 Jupiter and Mars orbit the Sun; their orbits are approximated as coplanar circles. The orbits of a 3 : 2 resonant asteroid that barely crosses Jupiter's orbit and a 4 : 1 resonant asteroid that barely crosses Mars's orbit are shown.

with radius a_{jup} and mean motion n_{jup}. We also take the liberty of circularizing the orbit of Mars and placing it in the same plane as Jupiter's orbit. The orbit of Mars has a radius $a_{\mathrm{mars}} \approx 0.293 a_{\mathrm{jup}}$ and thus a mean motion $n_{\mathrm{mars}} \approx n_{\mathrm{jup}}/(0.293)^{3/2} \approx 6.31 n_{\mathrm{jup}}$. Now we add an infinitesimal test mass, which acts as our asteroid. The orbital motion of the asteroid is coplanar with the Sun–Jupiter and Sun–Mars orbits. (Thus, our problem is similar to the restricted three-body problem, but with the addition of Mars as an added gravitational hazard for the test-mass asteroid.) Now we assert that the asteroid orbits the Sun in the same sense as Jupiter and Mars (counterclockwise in Figure 3.2), on an orbit with $a_{\mathrm{mars}} < a_{\mathrm{ast}} < a_{\mathrm{jup}}$. The mean motion of the asteroid, n_{ast}, has a resonance with that of Jupiter, in the ratio

$$\frac{n_{\mathrm{ast}}}{n_{\mathrm{jup}}} = \frac{\alpha}{\beta}, \tag{3.2}$$

where α and β are mutually prime positive integers. Since $a_{\mathrm{ast}} < a_{\mathrm{jup}}$, this implies $\alpha > \beta$. If the asteroid's orbit has a non-zero eccentricity e_{ast}, then its distance from the Sun varies from

$$r_{\mathrm{pe}} = \left(\frac{\beta}{\alpha}\right)^{2/3} (1 - e_{\mathrm{ast}}) a_{\mathrm{jup}} \tag{3.3}$$

at periapsis to

$$r_{\text{ap}} = \left(\frac{\beta}{\alpha}\right)^{2/3} (1 + e_{\text{ast}}) a_{\text{jup}} \qquad (3.4)$$

at apoapsis.

Equation 3.4 illustrates one potential problem for the orbit of our resonant asteroid. If its eccentricity is greater than a critical value,

$$e_{\text{JX}} = \left(\frac{\alpha}{\beta}\right)^{2/3} - 1, \qquad (3.5)$$

then the asteroid is a "Jupiter-crosser"; that is, its periapsis is inside Jupiter's orbit, but its apoapsis is outside. For the hilda group of asteroids, with $\alpha/\beta = 3/2$, this critical eccentricity is $e_{\text{JX}} = 0.310$. Figure 3.2 shows the orbit of a hilda with the critical Jupiter-crossing eccentricity. Although no known hildas have $e > e_{\text{JX}}$, some of them have $e \sim 0.3$. Hildas with an eccentricity this large come very close to Jupiter's orbit; however, a plot of the hildas' location at any given time reveals that none of them are close to Jupiter itself. Consider the hildas at the far left of Figure 3.3. These objects are at apoapsis when Jupiter is on the far side of its orbit, 180° away as seen from the Sun. During a time interval $\Delta t = (2/3)\mathcal{P}_{\text{jup}} \approx 7.9\,\text{yr}$, this particular gang of hildas orbits through 360° and is back at apoapsis. During the same time, Jupiter orbits through only 240°, and now leads the gang of hildas by just 60° rather than

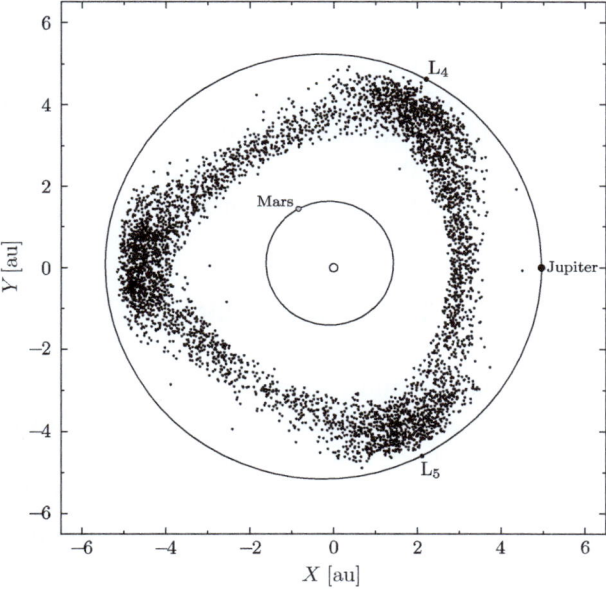

Figure 3.3 Location of the hildas (black points) relative to Jupiter, Mars, and the Jupiter–Sun L₄ and L₅ Lagrange points. Plotted at the same epoch as Figure 2.4.

$180°$. After another interval $\Delta t = (2/3)\mathcal{P}_{jup}$, the gang of hildas is back at apoapsis, but Jupiter now trails the gang by $60°$. This resonance produces the "hilda triangle" shown in Figure 3.3 and ensures that when any hilda sees Jupiter in opposition, the hilda will be at periapsis.

For asteroids with large values of α/β, crossing the orbit of Mars poses a greater hazard than crossing the orbit of Jupiter. From Equation 3.4, we find that asteroids with $\alpha/\beta > 2^{3/2} \approx 2.83$ never cross Jupiter's orbit, even in the limit $e_{ast} \rightarrow 1$. However, Equation 3.3 tells us that an asteroid with any value of α/β can be a Mars-crosser, as long as its eccentricity is greater than a critical value,

$$e_{MX} = 1 - \frac{a_{mars}}{a_{jup}} \left(\frac{\alpha}{\beta} \right)^{2/3} \approx 1 - 0.293 \left(\frac{\alpha}{\beta} \right)^{2/3}. \tag{3.6}$$

If we place our test-mass asteroid in the $4:1$ Kirkwood gap, the eccentricity required for a Mars-crossing orbit has the fairly modest value $e_{MX} = 0.262$; Figure 3.2 shows the orbit of a $4:1$ resonant asteroid that is just barely a Mars-crosser. Although this Mars-crosser is in resonance with Jupiter, it is not in resonance with Mars. Thus, each time it is at periapsis, and is located at the orbit of Mars, the planet Mars itself will be at a different point on its orbit. Eventually, the Mars-crosser will have a very close encounter with Mars that will fling it onto an orbit with a different semimajor axis.

Although Mars provides a mechanism for removing high-eccentricity asteroids from the Kirkwood gaps, this mechanism also works for high-eccentricity asteroids that are not in resonance with Jupiter. It also fails to explain why the Kirkwood gaps are empty of asteroids with $e_{ast} < e_{MX}$; there are plenty of low-eccentricity asteroids outside the Kirkwood gaps. Thus, there must be an additional mechanism that takes asteroids at mean motion resonances with Jupiter and increases their eccentricity until they are Mars-crossers. This mechanism involves the concept of chaotic orbits.

3.2 Chaotic Orbits

Isaac Newton showed that a bound system of two spherical objects has elliptical orbits with fixed a and $\mathcal{P}$. However, he was also aware that in the multi-body solar system, there would be perturbations from "the mutual Actions of Comets and Planets upon one another." Pierre-Simon Laplace, in the early nineteenth century, examined whether the gravitational forces between planets would ultimately destabilize the solar system. Although Laplace was limited to analytic approximations, in his monumental work *Mécanique Céleste*, he was able to show that the solar system is stable on short time spans. Given this stability, Laplace developed a highly deterministic world view; once the initial conditions of the universe were set up, in Laplace's view, everything would run in a perfectly predetermined manner. In 1814, he wrote of "an intellect which at a certain moment could know all forces that set nature in motion, and all positions of all items of which nature is composed." Laplace concluded that "for such an intellect, nothing would be uncertain and the future just like the past would be present before its eyes."

Even before the coming of quantum mechanics, this deterministic view of the universe was found to have practical flaws. The study of chaos was inadvertently boosted by King Oscar II of Sweden in the 1880s, when he offered a prize of 2500 kronor for a solution of the restricted three-body problem.[2] The mathematician Henri Poincaré, in his resulting studies of the restricted three-body problem, found that this very simple system displayed chaotic behavior. The modern definition of **deterministic chaos** is the study of deterministic systems (that is, systems in which the behavior is completely determined by the initial conditions) that yield widely diverging outcomes for extremely tiny changes in the initial conditions. As summed up by Poincaré, "It may happen that small differences in the initial conditions produce very great ones in the final phenomena. A small error in the former will produce an enormous error in the latter" (1911, p. 94).

In Section 2.1, our discussion of the restricted three-body problem did not discuss chaotic effects. How can we determine which orbits of a test mass are chaotic, and which are not? Let's use the Sun and Jupiter as our primary and secondary, and consider a simulated asteroid that we start on a heliocentric orbit with semimajor axis $a_i = 0.6944a_{\text{jup}}$ and eccentricity $e_i = 0.2065$. This orbit is close to the $7:4$ inner mean motion resonance with Jupiter, which is at $a = (4/7)^{2/3}a_{\text{jup}} = 0.6886a_{\text{jup}}$. However, the initial eccentricity is much smaller than the Jupiter-crossing eccentricity $e_{\text{JX}} \approx 0.44$ for an orbit of this size. Thus, Jupiter has only a small perturbative effect on the orbit of this asteroid. After several orbits, the simulated asteroid settles into an orbit whose semimajor axis and eccentricity are shown in Figure 3.4. Although a and e change with time, the eccentricity stays in the range $e = 0.238(1 \pm 0.12)$ over the course of 300 Jupiter orbits, corresponding to $\sim$520 asteroid orbits. Similarly, the orbit size stays in the range $a/a_{\text{jup}} = 0.687(1 \pm 0.02)$. Although the properties of this asteroid's orbit are not perfectly periodic, they can be validly described as "quasi-periodic." Figure 3.4 also illustrates that for a test mass near a mean motion resonance with a massive secondary there will be variation in the value of a, and thus $\mathcal{P}$, for its orbit. (Thus, it is valid to say that Pluto is in a $3:2$ outer resonance with Neptune, even though the current value of $n_{\text{N}}/n_{\text{P}}$ is not exactly 1.5.)

---

[2] In those days, a krona was worth something; Sweden was on the gold standard, with 2500 kronor equaling 1008 grams of pure gold.

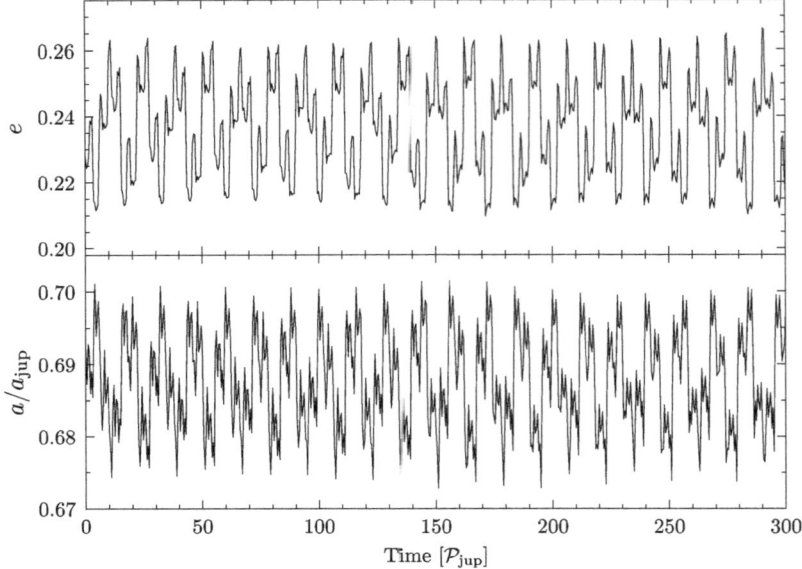

Figure 3.4 Eccentricity (upper panel) and semimajor axis (lower panel) for an aster-oid with a quasi-periodic orbit. Initial orbital elements for the simulation were $a_i = 0.6944a_{jup}$ and $e_i = 0.2065$. [Calculated using REBOUND following Murray and Dermott 1999]

Figure 3.4 gives a visual impression of quasi-periodic, rather than chaotic, variations. However, Poincaré has given us a more rigorous tool for identify-ing chaotic orbits: the **surface of section**. To determine the future trajectory of our test-mass asteroid, we need to know, at some time t, its location (X, Y) and velocity $(\dot{X}, \dot{Y})$ in the rotating frame of reference. That is, we must know its position in a four-dimensional **phase space** $(X, Y, \dot{X}, \dot{Y})$. The test mass cannot wander freely through all four dimensions; it is restricted to the three-dimensional "surface" defined by the Jacobi integral C_J (Equation 2.14). Visualizing a four-dimensional space is rather difficult, but Poincaré helped out by taking useful two-dimensional slices of the four-dimensional phase space. Poincaré's idea was to make a plot whose axes were X and $\dot{X}$. He then placed a dot at the $(X, \dot{X})$ location of the test mass whenever it was located on the $Y = 0$ axis (that is, on the line defined by the primary and secondary) and had $\dot{Y} > 0$.

As an example, Figure 3.5 shows the surface of section for the quasi-periodic orbit whose eccentricity and semimajor axis are plotted in Figure 3.4. The surface of section shows three distinct isolated "islands." In general, for non-chaotic orbits near the $\alpha : \beta$ resonance (where α and β are mutually prime integers with $\alpha > \beta$) the surface of section shows $n_{is} = \alpha - \beta$ islands. Each time the test mass passes through the $Y = 0$ line going in the positive Y

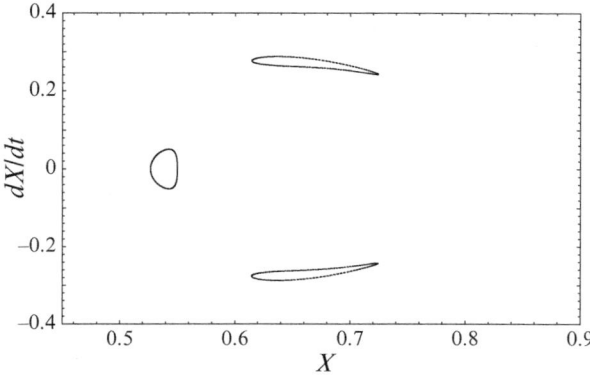

Figure 3.5 Surface of section for the quasi-periodic orbit examined in Figure 3.4. Distance is in units of a_{jup}. [Murray and Dermott 1999]

direction, a dot appears in one of the islands, with all n_{is} islands being visited in turn.

Now let's put our test-mass asteroid on a slightly different orbit near the 7 : 4 resonance. In Figure 3.6, the initial semimajor axis was chosen to be $a_i =$ $0.6984a_{jup}$, just 0.6% larger than the orbit assumed in Figure 3.4. In addition, the initial eccentricity $e_i \sim 0.2$ is very similar for both orbits. Figure 3.6 shows

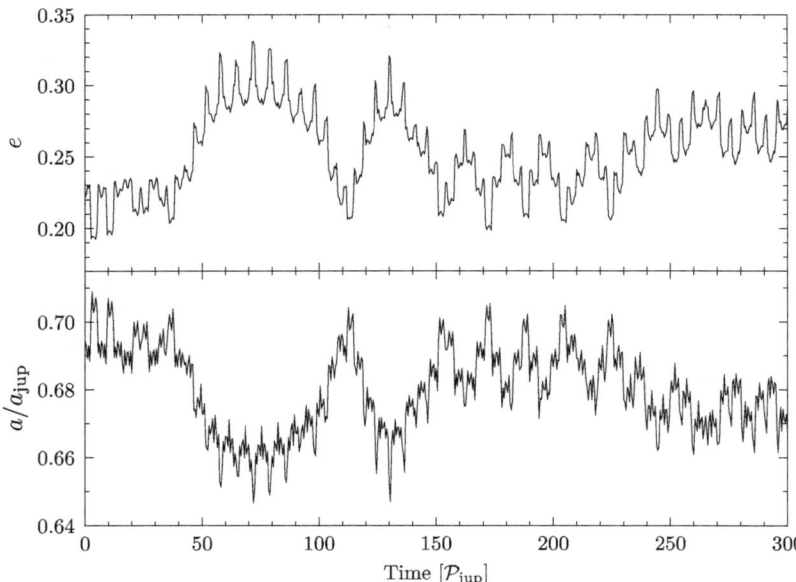

Figure 3.6 Eccentricity (upper panel) and semimajor axis (lower panel) versus time for an asteroid on a chaotic orbit. Initial orbital elements for the simulation were $a_i = 0.6984a_{jup}$ and $e_i = 0.1967$. [Calculated using REBOUND following Murray and Dermott 1999]

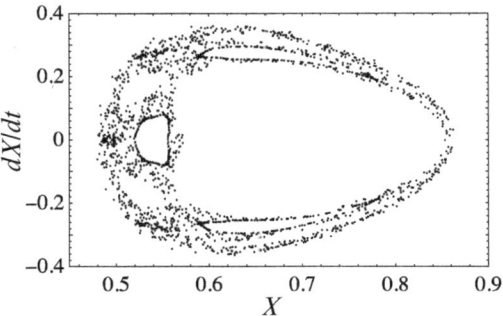

Figure 3.7 Surface of section for the chaotic orbit examined in Figure 3.6. Distance is in units of a_{jup}. [Murray and Dermott 1999]

behavior characteristic of chaotic orbits. Although the eccentricity starts by wavering around the value $e \approx 0.22$, it later jumps unpredictably to values near $e \approx 0.30$. At $t \approx 150\mathcal{P}_{\mathrm{jup}}$, it seems to settle into quasi-periodic behavior, but then jumps to larger e and smaller a. The surface of section for this orbit (Figure 3.7) shows behavior typical of chaotic orbits. Although it shows patterns similar to those of the non-chaotic orbit, it lacks isolated, well-defined "islands." However, remember that "chaos" does not imply the abandonment of all rules. Chaotic orbits in the restricted three-body problem, just like non-chaotic orbits, must conserve the Jacobi integral.

Chaotic orbits play a major role in emptying the Kirkwood gaps. Numerical studies shown that there is a concentration of chaotic orbits in the vicinity of the mean motion resonances with Jupiter. If an asteroid is on such a chaotic orbit, then (as seen in Figure 3.6) an increase in the eccentricity of a chaotic orbit is accompanied by a decrease in its semimajor axis. Thus, the apoapsis distance $r_{\mathrm{ap}} = a(1 + e)$ of the chaotic asteroid remains roughly constant; for the asteroid in Figure 3.6, we find $r_{\mathrm{ap}} \approx 0.85a_{\mathrm{jup}}$, keeping the asteroid at a safe distance from Jupiter. However, the periapsis distance $r_{\mathrm{pe}} = a(1-e)$ of the chaotic asteroid varies significantly; for the asteroid in Figure 3.6, the periapsis distance wanders over the range $r_{\mathrm{pe}} = (0.43 \rightarrow 0.59)a_{\mathrm{jup}}$ during the time interval examined. If a chaotic asteroid has its periapsis distance driven below $a_{\mathrm{mars}} = 0.293a_{\mathrm{jup}}$, then the asteroid is a Mars-crosser, and is liable to have its semimajor axis greatly changed by a close encounter with Mars, and thus be kicked away from the original mean motion resonance.

3.3 Asteroid Families

A histogram of the semimajor axes of asteroids (Figure 3.1) shows the presence of the Kirkwood gaps, related to mean motion resonances with Jupiter.

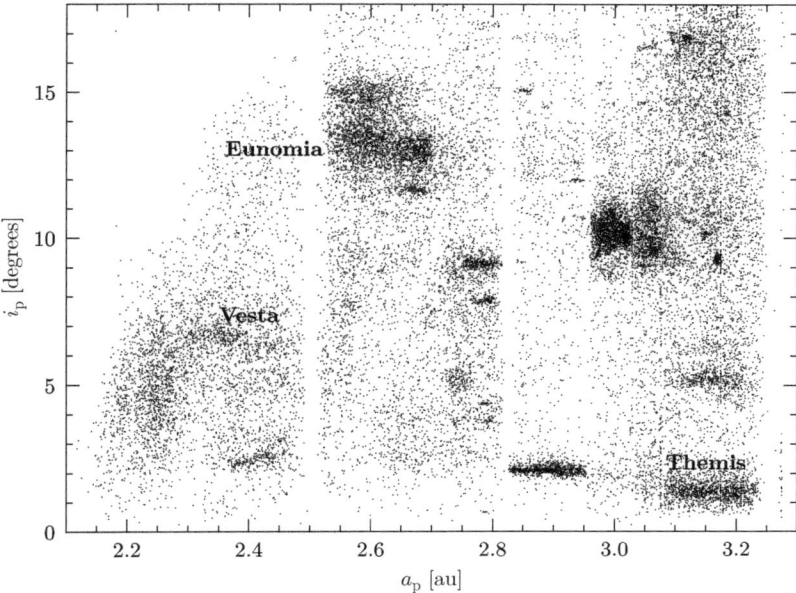

Figure 3.8 Proper orbital elements (inclination versus semimajor axis) for ∼34 000 asteroids in the main belt with diameters ∼5 km or greater. The asteroid families discussed in the text are labeled.

However, we can add additional information by looking at the inclination i of each asteroid's orbit relative to the ecliptic plane.[3] A plot of inclination versus semimajor axis for asteroid orbits (Figure 3.8) reveals concentrations of asteroids at certain preferred combinations of a and i. These concentrations are called **asteroid families**. The members of an asteroid family are found to have similar values of eccentricity e as well as having similar values of a and i. Asteroid families are named after the most prominent member of the family. For example, there is the Vesta family ($a \sim 2.4$ au, $i \sim 7°$, $e \sim 0.1$), the high-inclination Eunomia family ($a \sim 2.6$ au, $i \sim 13°$, $e \sim 0.15$), and the low-inclination Themis family ($a \sim 3.1$ au, $i \sim 1.5°$, $e \sim 0.15$).

One difficulty in identifying asteroid families is that an asteroid's values for a, i, and e are constantly changing. Figures 3.4 and 3.6 show the changes in a and e for an asteroid strictly confined to Jupiter's orbital plane; similar numerical studies of inclined orbits reveal that i is also changed by the perturbing effect of Jupiter. To a lesser extent, other planets such as Saturn and Mars also cause variations in a, e, and i. At any time t, an asteroid has a location $\vec{x}(t)$ and velocity $\vec{v}(t)$ relative to the barycenter of the solar system. Plots such as Figures 3.4 and 3.6 show what are called the **osculating** orbital elements. These are

---

[3] Although the ecliptic plane is defined by the Earth's orbit around the Sun, the inclination of Jupiter's orbit relative to the Earth's is only 1.3°.

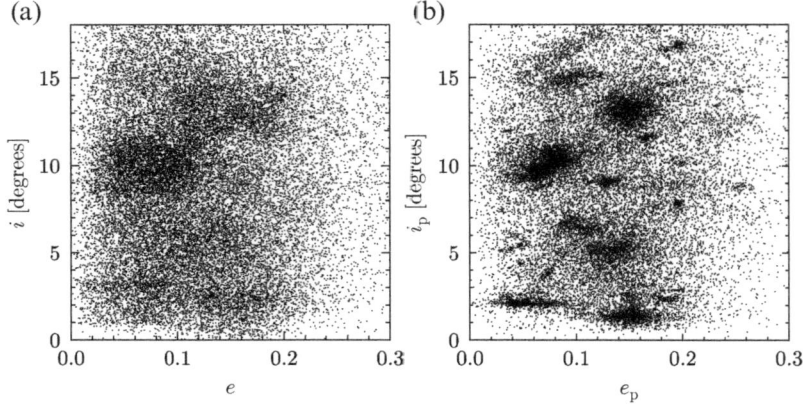

Figure 3.9 (a) Osculating inclination versus osculating eccentricity for the asteroid sample of Figure 3.8. (b) A similar plot for the same sample using proper inclination and proper eccentricity.

the orbital elements $a(t)$, $e(t)$, and $i(t)$ of the perfect Keplerian orbit that passes through the asteroid's location $[\vec{x}(t), \vec{v}(t)]$ in six-dimensional phase space.[4]

If the osculating elements vary in a quasi-periodic manner, then averaging their value over a long timescale recovers the **proper** orbital elements a_p, i_p, and e_p. The difference between an osculating orbital element and the corresponding proper orbital element is called the forced orbital element. In the case of asteroids, the forced orbital element is primarily the result of Jupiter's gravitational perturbations. For most asteroids the forced element has an amplitude $\Delta a < 0.02$ au, $\Delta i < 2°$, and $\Delta e < 0.1$. As an example, the large asteroid 4 Vesta, head of the Vesta family, has proper orbital elements $a_p = 2.361\,51$ au, $i_p = 6.392\,34°$, and $e_p = 0.098\,758$. By comparison, the osculating values on Julian date[5] $2\,460\,000.5$ were $a = 2.363\,04$ au, $i = 7.139\,26°$, and $e = 0.0887\,58$.

At any given instant, a plot of osculating inclination versus osculating eccentricity, as in Figure 3.9(a), is not perfectly smooth; you can see that some regions of the i versus e plane are more popular with asteroids. However, the existence of asteroid families doesn't pop out vividly until proper inclination is plotted versus proper eccentricity, as in Figure 3.9(b). When Kiyotsugu Hirayama discovered the existence of asteroid families in 1918, he did so using proper orbital elements.

Once an asteroid family is identified from the orbital properties of its members, it is also found that the family members have similar spectra of reflected

---

[4] The word "osculating" comes from the Latin *osculum*, meaning "kiss." Osculating orbital elements are those of a Keplerian orbit that barely kisses the asteroid's true orbit at time t.

[5] This date corresponds to either 2023 Feb 24 or 2023 Feb 25, depending on whether you were west or east of Greenwich at the time.

sunlight. Members of the Eunomia family all have S-type spectra, characteristic of stony (or siliceous) material. Members of the Themis family, farther out in the asteroid belt, have C-type spectra, characteristic of dark carbonaceous material. Most members of the Vesta family have V-type spectra, with spectra similar to that of 4 Vesta itself. The "hildas," incidentally, comprise at least two asteroid families that happen to live in the same neighborhood at $a_p \approx 3.97$ au; the Schubart family has $e_p \approx 0.2$ and $i_p \approx 3°$ while the Hilda family has a similar eccentricity but a larger inclination, with $i_p \approx 9°$.

Members of an asteroid family have similar orbital elements and similar surface properties since they result from the collisional breakup of a single large asteroid. It is estimated that 24 Themis contains $\sim15\%$ of the mass of the original parent of the Themis family, that 15 Eunomia contains $\sim70\%$ of the mass of the parent of the Eunomia family, and that 4 Vesta still contains 98% of the mass of the parent of the Vesta family. It appears that the collision that created the Vesta family was the same collision that gouged out the large impact basin Rheasilvia in the southern hemisphere of Vesta; geological evidence (or should that be vestalogical evidence?) indicates that the collision occurred ~1 Gyr ago.

The age of the Themis and Eunomia families can be estimated by making use of the **Yarkovsky effect**. Although in this book we concentrate mainly on gravitational forces, there are circumstances in which non-gravitational effects, such as the Yarkovsky effect, can measurably change an object's orbit. Briefly, the Yarkovsky effect results from the non-uniform temperature of an asteroid's surface and the resulting anisotropic thermal radiation. Imagine a rotating asteroid whose sense of rotation is the same as its sense of orbital motion (that is, *prograde* rotation). The coldest region on the asteroid's surface, given a uniform albedo over the surface, is where the local time is just before sunrise; this is on the leading hemisphere of the asteroid. Conversely, the hottest region is where it is locally late afternoon; this is on the trailing hemisphere. Thus, the flux of thermal radiation is greater from the trailing hemisphere than from the leading hemisphere. The resulting radiative acceleration is in the general direction of the asteroid's orbital motion, increasing the angular momentum of the asteroid's orbit, and causing its semimajor axis a to increase. Since radiative force is proportional to the asteroid's surface area ($F \propto d^2$) and the asteroid's mass is proportional to volume ($m \propto d^3$), the Yarkovsky acceleration ($F/m \propto d^{-1}$) is greater for smaller asteroids.

The size of the Yarkovsky acceleration is not large for a midsize asteroid. For example, the asteroid 6489 Golevka, an irregular Earth-crosser with mean diameter $d \approx 0.53$ km, has an orbit that is steadily decreasing in size, implying it's a retrograde rotator. The semimajor axis of the orbit is currently shrinking at the rate of 3 μm s^{-1}. Although a snail can ooze along at 1000 times this speed, it does add up over the age of the solar system, with

$da_p/dt \sim -3\,\mu\mathrm{m\,s^{-1}} \sim -0.6\,\mathrm{au\,Gyr^{-1}}$. The magnitude of the Yarkovsky effect depends on many variables; however, we expect that all members of an asteroid family will have a very similar albedo, and start at the same distance from the Sun. This means that the magnitude of the Yarkovsky effect will depend mainly on an asteroid's mean diameter d and rotation rate ω. As time goes on, family members with the same diameter d will develop a spread in semimajor axes, with prograde rotators having $da_p/dt > 0$ and retrograde rotators having $da_p/dt < 0$. Given the known albedos, diameters, and spread in a_p for the Themis and Eunomia families, it has probably been $\sim 2\,\mathrm{Gyr}$ since the collisions that created them. The members of an asteroid family, we conclude, can "remember" their proper orbital elements a_p, i_p, and e_p for a good fraction of the age of the solar system.

3.4 Close Encounters

As long as an asteroid doesn't land in a chaotic orbit near a mean motion resonance with Jupiter, its life tends to be stable. The perturbative effects of Jupiter and other planets cause quasi-periodic fluctuations of a, i, and e around their proper values, while the proper semimajor axis increases only gradually with time because of the Yarkovsky effect. What happens, however, to those asteroids whose chaotic orbits bring them very close to Jupiter? For that matter, what happens to a comet when its eccentric orbit happens to bring it close to Jupiter? First, let's assume that the asteroid or comet doesn't physically collide with Jupiter, and that it doesn't come close enough to be tidally disrupted. Although the semimajor axis, inclination, and eccentricity of the object can all be greatly altered by a close gravitational encounter, there is one constant of integration that will not change. Since the mass of a comet is negligible compared to that of Jupiter, and since Jupiter's orbit is close to circular, we can approximate the Sun–Jupiter–comet system as a restricted three-body problem.[6] Since the orbit of a comet generally has a large inclination, we can't assume a *coplanar* restricted three-body problem. However, the Jacobi integral,

$$C_J = -\dot{X}^2 - \dot{Y}^2 - \dot{Z}^2 + n^2(X^2 + Y^2) + 2\mu\left(\frac{1-f}{r_1} + \frac{f}{r_2}\right), \qquad (3.7)$$

is still conserved as the comet goes through its close encounter with Jupiter. (Compare to Equation 2.14 for the Jacobi integral when the test mass is confined to a plane.) In the specific case of the Sun–Jupiter system, the standard gravitational parameter is $\mu = 1.3284 \times 10^{26}\,\mathrm{cm^3\,s^{-2}}$, the fraction

---

[6] We're calling the test mass a "comet" in this problem, but the same physics applies to asteroids and to derelict interplanetary spacecraft.

of mass contained by Jupiter is $f = 9.5368 \times 10^{-4}$, and the mean motion of the Sun–Jupiter binary is $n = \mu^{1/2}/a_{12}^{3/2} = 1.6780 \times 10^{-8}\,\text{s}^{-1}$.

In an inertial coordinate system whose origin is the Sun–Jupiter barycenter and whose z axis coincides with the Z axis of the corotating frame, the coordinate transformation is

$$X = x \cos nt + y \sin nt, \tag{3.8}$$

$$Y = -x \sin nt + y \cos nt, \tag{3.9}$$

$$Z = z. \tag{3.10}$$

The time $t = 0$ is chosen to be an instant when the spinning X axis coincides with the inertial x axis. In terms of the inertial coordinates, the Jacobi integral of the comet is

$$C_J = -\dot{x}^2 - \dot{y}^2 - \dot{z}^2 + 2n^2(x\dot{y} - y\dot{x}) + 2\mu\left(\frac{1-f}{r_1} + \frac{f}{r_2}\right). \tag{3.11}$$

Using the relation $n^2 = \mu/a_{12}^3$, this can be rewritten as

$$C_J = -\dot{x}^2 - \dot{y}^2 - \dot{z}^2 + \frac{2\mu}{a_{12}^3}(x\dot{y} - y\dot{x}) + 2\mu\left(\frac{1-f}{r_1} + \frac{f}{r_2}\right). \tag{3.12}$$

Equation 3.12 is an exact result for an infinitesimal test mass, regardless of where it is relative to the Sun and Jupiter. However, we can now make some useful approximations based on the fact that $f \ll 1$ for the Sun–Jupiter system. When we look at the gravitational potential term,

$$\Phi = 2\mu\left(\frac{1-f}{r_1} + \frac{f}{r_2}\right), \tag{3.13}$$

we note that as long as the comet stays far from Jupiter, with $r_2 \gg fr_1$, the potential it feels from Jupiter is negligible compared to that of the Sun. In addition, when we look at the Sun's potential, we can set $1 - f \approx 1$ and $r_1 \approx r$, thus ignoring the relatively small offset fa_{12} of the Sun–Jupiter barycenter from the Sun's center. With these approximations, the Jacobi integral of Equation 3.12 becomes

$$C_J \approx -\dot{x}^2 - \dot{y}^2 - \dot{z}^2 + \frac{2\mu}{a_{12}^3}(x\dot{y} - y\dot{x}) + \frac{2\mu}{r}, \tag{3.14}$$

where r is the distance of the comet from the Sun–Jupiter barycenter.

The comet's initial orbit around the Sun–Jupiter barycenter has semimajor axis a_i, inclination i_i, and eccentricity e_i. For an elliptical orbit, the comet's speed is given by the *vis viva* equation (Equation 1.52):

$$\dot{x}^2 + \dot{y}^2 + \dot{z}^2 \approx \mu\left(\frac{2}{r} - \frac{1}{a_i}\right). \tag{3.15}$$

Another conserved quantity on an elliptical orbit is the specific angular momentum (compare to Equation 2.52),

$$j = \mu^{1/2} a_i^{1/2} (1 - e_i^2)^{1/2}. \tag{3.16}$$

Since the plane of the comet's initial orbit is constant, so is the component of j in the z direction:

$$j_z = x\dot{y} - y\dot{x} = \mu^{1/2} a_i^{1/2} (1 - e_i^2)^{1/2} \cos i_i. \tag{3.17}$$

Substituting Equations 3.15 and 3.17 into the Jacobi integral of Equation 3.12, we find that the Jacobi integral of the comet on its initial orbit, far from the perturbations of Jupiter, is

$$C_{J,i} \approx \frac{\mu}{a_i} + \frac{2\mu}{a_{12}} \left(\frac{a_i}{a_{12}} \right)^{1/2} (1 - e_i^2)^{1/2} \cos i_i. \tag{3.18}$$

Suppose the comet now has a close encounter with Jupiter. If the encounter doesn't fling the comet out of the solar system entirely, it will settle back onto an elliptical orbit around the Sun, with new orbital elements a_f, i_f, and e_f. The new Jacobi integral will then be

$$C_{J,f} \approx \frac{\mu}{a_f} + \frac{2\mu}{a_{12}} \left(\frac{a_z}{a_{12}} \right)^{1/2} (1 - e_f^2)^{1/2} \cos i_f. \tag{3.19}$$

Since the constancy of the Jacobi integral is something we can cling to in the restricted three-body problem (coplanar or not), we can set Equation 3.18 equal to Equation 3.19, and find

$$\frac{a_{12}}{a_i} + 2 \left(\frac{a_i}{a_{12}} \right)^{1/2} (1 - e_i^2)^{1/2} \cos i_i = \frac{a_{12}}{a_f} + 2 \left(\frac{a_f}{a_{12}} \right)^{1/2} (1 - e_f^2)^{1/2} \cos i_f. \tag{3.20}$$

This equality is known as **Tisserand's criterion**, after the astronomer Félix Tisserand.

Tisserand's criterion was originally thought of as a way to determine whether a newly discovered comet was actually the return of a previously known comet that had undergone a close encounter with Jupiter. Such an encounter changes the orbital elements of the comet, such as the semimajor axis a, eccentricity e, and inclination i. However, it will leave unchanged the comet's **Tisserand parameter**

$$T(a, e, i) = \frac{a_{12}}{a} + 2 \left(\frac{a}{a_{12}} \right)^{1/2} (1 - e^2)^{1/2} \cos i. \tag{3.21}$$

Remember from Section 2.4 that for ZLK oscillations, in which the massive perturbing object always remains distant, the values of a and $f_z = (1 - e^2)^{1/2} \cos i$ are separately conserved. For close encounters of a test mass with a perturbing object, only the combination $T(a, f_z)$ is conserved.

Note that for a given cometary orbit, the Jovian Tisserand parameter,

$$T_{jup} = \frac{a_{jup}}{a} + 2 \left(\frac{a}{a_{jup}} \right)^{1/2} (1 - e^2)^{1/2} \cos i_{jup}, \tag{3.22}$$

where a_{jup} = 5.20 au, is different from the Saturnian Tisserand parameter,

$$T_{sat} = \frac{a_{sat}}{a} + 2 \left(\frac{a}{a_{sat}} \right)^{1/2} (1 - e^2)^{1/2} \cos i_{sat}, \tag{3.23}$$

where a_{sat} = 9.57 au. (The orbits of Jupiter and Saturn have a mutual inclination of 1.2°, so i_{jup} and i_{sat} are similar but not identical for a given cometary orbit.) The Tisserand parameters for other planets will be different as well. Thus, not only does the Tisserand parameter help to identify comets that have undergone a close encounter with a planet, it points a finger at the planet responsible.

More recently, the Tisserand parameter has been used to distinguish between different classes of astronomical objects. Asteroids and cometary nuclei cannot be cleanly separated on the basis of the light they reflect; some classes of asteroid have reflectance spectra similar to that of a comet's nucleus. However, a plot of the Jovian Tisserand parameter for asteroids and comets, as shown in Figure 3.10, reveals that asteroids generally have a larger Tisserand parameter than comets at the same distance from the Sun. An asteroid with a semimajor axis a that places it in the main asteroid belt generally has low eccentricity and inclination, as shown in Figure 3.9. Because of their relatively low e and i_{jup}, asteroids usually have a Jovian Tisserand parameter that is within a few percent of the maximum Tisserand parameter for their semimajor axis:

$$T_{jup,max}(a) = \frac{a_{jup}}{a} + 2 \left(\frac{a}{a_{jup}} \right)^{1/2}. \tag{3.24}$$

This means that in the main asteroid belt, nearly all asteroids have $3 < T_{jup} < 3.7$. However, comets (shown as the gray dots in Figure 3.10) typically have $T_{jup} < 3$. Comets on retrograde orbits, with $i_{jup} > 90°$, can even have $T_{jup} < 0$, if their orbit is sufficiently large. For instance, Comet Halley has $T_{jup} \approx -0.6$.

Dividing comets from asteroids by their Jovian Tisserand parameter is not foolproof. For instance, the two gray dots in the upper left corner of Figure 3.10, indicating "comets" with $T_{jup} \sim 3.6$, represent the objects 354P/LINEAR and 311P/PanSTARRS, which were originally designated as comets because they had visible tails. However, closer examination revealed that they may be rubble-pile asteroids emitting dust because of a recent collision (354P/LINEAR) or rapid rotation (311P/PanSTARRS).

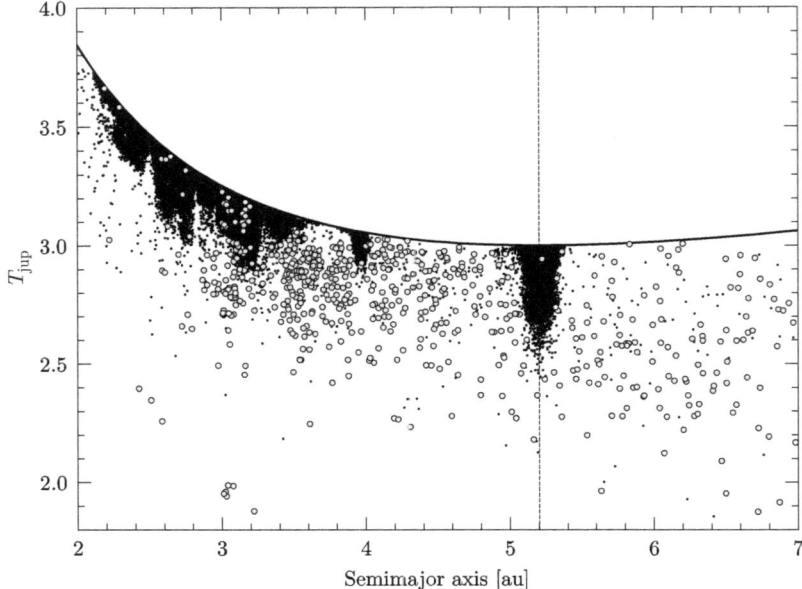

Figure 3.10 Jovian Tisserand parameter T_{jup} for asteroids (black points) and comets (gray dots). The vertical dashed line indicates the Sun–Jupiter distance, $a_{\mathrm{jup}} = 5.20\,\mathrm{au}$; the curved solid line is $T_{\mathrm{jup,max}}$.

3.5 Laplace-Like Resonances

In studying the dynamics of asteroids, we safely assumed that they are infinitesimal test masses, and do not have a significant gravitational effect on each other. (As the impact basin on 4 Vesta testifies, asteroids do sometimes have physical collisions, but their gravitational effect on each other is negligible.) However, in systems with multiple moons of roughly the same mass orbiting the same planet, gravitational interactions between the moons can lead to intricate resonances among the moons. One long-studied example of a multiple-resonance system is given by the three inner Galilean moons of Jupiter: Io, Europa, and Ganymede. (The Galilean moons are well inside Jupiter's Hill radius, so we will ignore the effect of the Sun on their orbits around Jupiter.) The mean motion of Io around Jupiter is $n_{\mathrm{I}} = 3.551\,55\,\mathrm{d}^{-1}$, that of Europa is $n_{\mathrm{E}} = 1.769\,32\,\mathrm{d}^{-1}$, and that of Ganymede is $n_{\mathrm{G}} = 0.878\,21\,\mathrm{d}^{-1}$. Thus, the commensurabilities in this system are

$$\frac{n_{\mathrm{I}}}{n_{\mathrm{E}}} = 2.0073, \qquad\qquad \frac{n_{\mathrm{E}}}{n_{\mathrm{G}}} = 2.0147. \tag{3.25}$$

These are both close to, but not identical with, the ratio $2:1$.[7]

---

[7] A footnote for Callisto fans: The outermost Galilean moon, Callisto, has a different commensurability. Callisto's mean motion $n_{\mathrm{C}} = 0.376\,49\,\mathrm{d}^{-1}$ leads to $n_{\mathrm{G}}/n_{\mathrm{C}} = 2.3326$, which is close to, but not identical with, the ratio $7:3$.

The $4:2:1$ mean motion resonance among Io, Europa, and Ganymede is not exact. However, when we look at the differences between mean motions rather than the ratios, we find the exact relation (within observational accuracy)

$$n_{\text{I}} - n_{\text{E}} = 2(n_{\text{E}} - n_{\text{G}}) = 1.782\,23\,\text{d}^{-1}. \tag{3.26}$$

To see what this equality implies for the Io–Europa–Ganymede system, imagine that you are on Europa for a bit of ice fishing. From your viewpoint, there is an **inferior conjunction** of Io whenever Io makes its closest approach to the center of Jupiter's disk. (In the idealized case of perfectly circular coplanar orbits, inferior conjunction occurs when the center of Io lies on the line segment connecting the centers of Jupiter and Europa.) The synodic period $\mathcal{P}_{\text{IE}}$ of Io as seen from Europa is the time between successive inferior conjunctions, and is given by the relation

$$\frac{2\pi}{\mathcal{P}_{\text{IE}}} = n_{\text{I}} - n_{\text{E}}. \tag{3.27}$$

From your viewpoint on Europa, you see Ganymede in **opposition** whenever it is directly opposite the center of Jupiter's disk. (In the idealized case, opposition occurs when the center of Europa lies on the line segment connecting the centers of Jupiter and Ganymede.) The synodic period $\mathcal{P}_{\text{EG}}$ of Ganymede as seen from Europa is the time between successive oppositions, and is given by

$$\frac{2\pi}{\mathcal{P}_{\text{EG}}} = n_{\text{E}} - n_{\text{G}}. \tag{3.28}$$

If we rewrite Equation 3.26 in terms of the synodic periods, we find that

$$\frac{2\pi}{\mathcal{P}_{\text{IE}}} = 2\frac{2\pi}{\mathcal{P}_{\text{EG}}} \tag{3.29}$$

and thus that $\mathcal{P}_{\text{EG}} = 2\mathcal{P}_{\text{IE}} = 7.050\,93$ days. When three objects orbiting the same massive primary have this $2:1$ ratio of their synodic periods, they are referred to as having a **Laplace resonance**.

Figure 3.11 shows the effects of the Laplace resonance among Io, Europa, and Ganymede. Panel (a) shows that when Ganymede is in opposition as seen from Europa, the moon Io is on the far side of Jupiter. Similarly, panel (c) shows that when Ganymede is in opposition as seen from Io, Europa is in exile on the far side of Jupiter. In this way, the Laplace resonance ensures that there is never a "triple conjunction" of Io, Europa, and Ganymede, with all three moons aligned on the same side of Jupiter.

When Pierre-Simon Laplace discussed the orbital resonance of the Galilean moons in *Mécanique Céleste*, they were the only example known of what is now called a Laplace resonance. More recently, similar resonances between synodic periods have been found within our solar system, as well as in systems

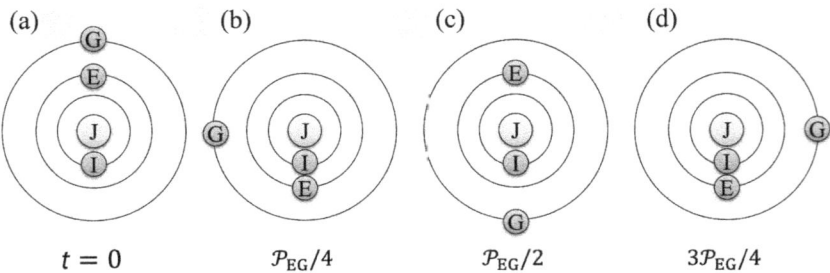

(a)	(b)	(c)	(d)
$t = 0$	$\mathcal{P}_{EG}/4$	$\mathcal{P}_{EG}/2$	$3\mathcal{P}_{EG}/4$

Figure 3.11 Four snapshots of Io (I), Europa (E), and Ganymede (G) orbiting the planet Jupiter (J). The images are separated in time by $\Delta t = \mathcal{P}_{EG}/4 = \mathcal{P}_{IE}/2 = 1.762\,73$ d. [Adapted from Paita *et al.* 2018]

of exoplanets orbiting other stars. Given the abundance of such systems, we need to extend the original definition of a Laplace resonance. Consider a system of three lower-mass objects orbiting a higher-mass primary; they have mean motions $n_1 > n_2 > n_3$. The three objects are in a **Laplace-like resonance** if the ratios n_1/n_2 and n_2/n_3 are approximately equal to the ratio of two small integers, and the ratio $(n_1 - n_2)/(n_2 - n_3)$ is exactly equal to the ratio of two small integers.[8]

As an example, the small moons Styx, Nix, and Hydra orbit the barycenter of Pluto and Charon. The mean motions of Styx and Nix have $n_1/n_2 \approx 5/4$, while the mean motions of Nix and Hydra have $n_2/n_3 \approx 3/2$. It is also found that

$$n_1 - n_2 = \frac{2}{3}(n_2 - n_3) = 0.0589\,\text{d}^{-1}. \tag{3.30}$$

This Laplace-like resonance implies that over the course of 213.4 days, an observer on Nix will see three oppositions of Hydra and two inferior conjunctions of Styx. As in the Io–Europa–Ganymede system, the phases of the orbits are such that there is never a triple conjunction of Nix, Styx, and Hydra all on the same side of Pluto.

In some exoplanetary systems, there are extended **resonance chains** of Laplace-like resonances. The TRAPPIST-1 system, for instance, consists of a red dwarf star orbited by seven known exoplanets, all on low-eccentricity, nearly coplanar orbits. The exoplanets are labeled TRAPPIST-1b through TRAPPIST-1h, in order of their distance from the star. (The letter "a" is reserved for the star itself.) Within this system, any three neighboring orbits obey a Laplace-like resonance. It is found that $n_b/n_c \approx 8/5$, $n_c/n_d \approx 5/3$, $n_d/n_e \approx 3/2$, $n_e/n_f \approx 3/2$, $n_f/n_g \approx 4/3$, and $n_g/n_h \approx 3/2$. In addition, to high accuracy, the mean motions obey the relations

---

[8] The short label "Laplace resonance" is now generally reserved for the special case that Laplace himself considered: $n_1/n_2 \approx 2$, $n_2/n_3 \approx 2$, and $(n_1 - n_2)/(n_3 - n_2) = 2$.

$$n_b - n_c = \frac{3}{2}(n_c - n_d) = 3(n_d - n_e) = \frac{9}{2}(n_e - n_f) \tag{3.31}$$

$$= 9(n_f - n_g) = 9(n_g - n_h) = 1.5645\,\mathrm{d}^{-1}.$$

It would be a suspicious coincidence if seven exoplanets happened by pure chance to have mean motions in a long resonance chain. However, if two exoplanets are not initially in resonance, and if one of the exoplanets drifts gradually inward or outward relative to the other, then if the two reach a resonance, they tend to be "locked" at that resonance.

While exoplanets are still in the process of formation, gas dynamics within the dusty, gaseous protoplanetary disk can cause them to drift inward or outward in semimajor axis. (Similar processes could have occurred in the circumplanetary disk from which the Galilean moons of Jupiter formed.) In addition, as we shall see in the next chapter, tidal effects can cause exoplanets and moons to gradually change their semimajor axis, long after the process of formation is over.

Exercises

3.1 Suppose an asteroid is in a $4:1$ mean motion resonance with Jupiter. The asteroid's orbit happens to have $i = 0°$ relative to the Earth's orbit.

(a) What is the minimum eccentricity e_{EX} required for this asteroid to be an Earth-crosser, with $r_{pe} = 1$ au?

(b) If the asteroid's orbit has $e = e_{EX}$, what is its orbital speed v_{pe} at periapsis?

(c) When the asteroid is at periapsis, it strikes the Earth. What is the relative speed of the asteroid and Earth just before impact?

(d) If the asteroid has a mass $m = 10^{10}$ g (corresponding to a rocky object with diameter $d \sim 20$ m), what is its kinetic energy just before impact?

3.2 Some comets enter the inner solar system on orbits with $e \approx 1$. In such a case, when you are not certain whether the comet is bound to the Sun ($e < 1$) or is an interstellar vagabond ($e > 1$), it is convenient to parameterize its orbit in terms of its periapsis distance $r_{pe} = a(1 - e)$ (where $a < 0$ for a hyperbolic orbit).

(a) Write the Jovian Tisserand parameter for a comet in terms of its orbital parameters r_{pe}, e, and i_{jup}.

(b) A comet enters the solar system on an initially parabolic orbit ($e_i = 1$), which is then altered by an encounter with Jupiter to become elliptical ($e_f < 1$). Write the Jovian Tisserand parameter in terms of the orbital parameters $r_{pe,i}$ and $i_{jup,i}$ of the initial parabolic orbit.

Then write the Jovian Tisserand parameter in terms of the orbital parameters a_f, e_f, and $i_{\mathrm{jup},f}$ of the final bound orbit.

(c) Consider the special case for which the comet's motion is prograde, and lies strictly in Jupiter's orbital plane. If the initial parabolic orbit has $r_{\mathrm{pe},i} = a_{\mathrm{jup}}$, what is the minimum possible value for the final eccentricity e_f, assuming the Tisserand parameter is conserved? What is a_f of this minimum eccentricity orbit? What are the corresponding perihelion and aphelion distances?

(d) Now consider the case in which the comet's motion is *retrograde,* and lies strictly in Jupiter's orbital plane. If the initial parabolic orbit has $r_{\mathrm{pe},i} = a_{\mathrm{jup}}$, show that the final orbit can be circular ($e_f = 0$) while still conserving the Tisserand parameter. What is the radius a_f of this circular retrograde orbit?

4

Tides

A Frenchman who arrives in London, will find Philosophy,
like every Thing else, very much chang'd there…
In France, 'tis the Pressure of the Moon that causes the Tides;
but in England 'tis the Sea that gravitates toward the Moon.

Letters Concerning the English Nation [1733]
Letter XIV: On Descartes and Sir Isaac Newton,
Voltaire (1694–1778)

So far, we have been considering the gravitational interaction between perfectly spherical bodies. In the case of many celestial objects, the "spherical cow" approximation is an excellent one. The ratio of the Sun's polar radius to its equatorial radius, for instance, is $R_{pol}/R_{eq} \approx 0.99999$. Although the Earth is slightly more flattened than the Sun, its axis ratio $R_{pol}/R_{eq} = 0.9966$ means it is still nearly spherical.

Some celestial objects, however, deviate significantly from a spherical shape. Asteroids and moons with mean diameter less than $d \sim 500$ km are usually irregular in shape, since their self-gravity is not large enough to overcome their internal compressional strength. Some asteroids are known to be extremely non-spherical. For instance, 1620 Geographos has a mean diameter $d \approx 2.8$ km, but radar imaging reveals it is an elongated, nearly prolate object with axis ratio $1:0.42:0.40$.[1] Turning our attention from solid asteroids to gaseous stars, it is seen that some young stars rotate very rapidly, with a rotation speed at their equator that approaches the escape speed from the star's surface. For these rapid rotators, the centrifugal term in their effective potential significantly flattens the star, making it into an oblate spheroid with $R_{pol}/R_{eq} \sim 0.7$ in the most extreme cases.

---

[1] For comparison, sports that use prolate balls favor axis ratios that range from $1:0.6:0.6$ (American football) to $1:0.7:0.7$ (Rugby League).

In a two-body system, as long as the periapsis distance between the two objects is much larger than the maximum diameter of each object, the deviations from an exact sphere cause only minor perturbations to their orbit.[2] However, when the distance between the two objects is no longer extremely large compared to the size of the larger object, deviations from spherical symmetry have a detectable effect on their orbit. Moreover, when the distance between the objects becomes this small, the objects will inevitably show deviations from spherical symmetry, due to the differential **tidal force** acting between the objects.

Here on Earth, the most obvious manifestation of tidal forces is the rise and fall of the Earth's liquid ocean relative to its stiff crust. Humans living near the ocean shore have long noted that the average time between high tides is 12 hours, 25 minutes, and that the average time between meridian transits of the Moon (that is, the time between one "lunar noon" and the next) is 24 hours, 50 minutes. Thus, it was long suspected that the Moon's influence caused the rise and fall of the tides. However, it wasn't until the development of Newtonian dynamics that a successful explanation for the tidal distortion of the Earth's ocean was developed. Even as late as the 1730s, as Voltaire jokingly pointed out, the (incorrect) explanation given by Descartes had not been completely defeated by the Newtonian explanation.

4.1 Tidal Forces

In the context of Newtonian dynamics, let's start our study of tides by looking at the effects of differential tidal forces in a two-body system. Both the primary and secondary in the system are spherical. For the moment, let's assume that the two objects are made of a rigid solid material. The primary has radius R_1 and mass m_1; the secondary has radius R_2 and mass $m_2 < m_1$. Now we bring the two objects close together; in Figure 4.1, as an example, the

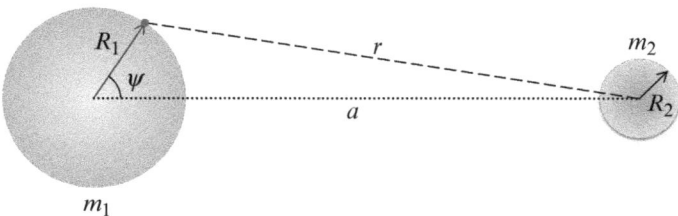

Figure 4.1 A secondary of mass m_2 exerts a gravitational force on a point mass located on the surface of a primary of radius R_1.

---

[2] Seen from a large enough distance, both a spherical cow and a non-spherical cow produce a gravitational potential indistinguishable from that of a point-mass cow with the same value of m_{cow}.

distance between the centers of the primary and secondary is $a = 6R_1$. If we pick a point on the surface of the primary, the gravitational potential at that point due to the presence of the spherical secondary is the Keplerian potential (Equation 1.13)

$$\Phi = -\frac{\mu_2}{r}, \tag{4.1}$$

where r is the distance from the chosen point to the center of the secondary. From the law of cosines,

$$r = a\left[1 - 2\left(\frac{R_1}{a}\right)\cos\psi + \left(\frac{R_1}{a}\right)^2\right]^{1/2}, \tag{4.2}$$

where the angle ψ, as shown in Figure 4.1, runs from $\psi = 0$ at the point closest to the secondary to $\psi = \pi$ at the antipodal point, farthest from the secondary. We can simply insert this value of $r(\psi)$ into Equation 4.1 if we want an exact solution for the potential Φ. However, in most cases of interest we can assume the orbital size a is much larger than the primary's radius R_1. In the limit that R_1/a is much smaller than one, we can usefully expand $1/r$ in terms of Legendre polynomials (compare to Equation 2.47):

$$\frac{1}{r} = \frac{1}{a}\sum_{\ell=0}^{\infty}\left(\frac{R_1}{a}\right)^{\ell}P_{\ell}(\cos\psi). \tag{4.3}$$

Keeping the first few terms in the expansion of $1/r$, we can write the gravitational potential Φ on the surface of the primary as

$$\Phi \approx -\frac{\mu_2}{a}\left[1 + \left(\frac{R_1}{a}\right)\cos\psi + \left(\frac{R_1}{a}\right)^2 P_2(\cos\psi)\right]. \tag{4.4}$$

The lowest-order ($\ell = 0$) term in the potential is thus

$$\Phi_{\ell=0} = -\frac{\mu_2}{a}. \tag{4.5}$$

This constant term does not produce an acceleration, and is thus not physically significant. The next ($\ell = 1$) term is

$$\Phi_{\ell=1} = -\frac{\mu_2}{a}\left(\frac{R_1}{a}\right)\cos\psi. \tag{4.6}$$

To the same ($\ell = 1$) order, the distance from our test point to the center of the secondary is (Equation 4.2)

$$r \approx a\left[1 - \left(\frac{R_1}{a}\right)\cos\psi\right]. \tag{4.7}$$

Thus, the $\ell = 1$ term in the potential increases linearly with distance from the secondary; it provides the acceleration $g = \mu_2/a^2$ required to keep the primary on a circular orbit around the barycenter.

It is the $\ell = 2$ term in Equation 4.4 that provides the tidal force:

$$\Phi_{\ell=2} \approx -\frac{\mu_2}{a} \left(\frac{R_1}{a} \right)^2 P_2(\cos \psi), \qquad (4.8)$$

where $P_2(\cos \psi) = (3 \cos^2 \psi - 1)/2$. To see how this potential can create **tidal bulges**, let's convert the primary to an "ocean world." That is, we take our rigid sphere of mass m_1 and radius R_1 and pour enough liquid on it to create a worldwide ocean. The mean depth d of the ocean is tiny compared to R_1, and its mass is tiny compared to m_1. We add this effectively massless ocean to the primary because, since the ocean is fluid, its surface must correspond to an equipotential surface of the gravitational potential. In the absence of the tide-creating secondary, the relevant potential would be due entirely to the primary:

$$\Phi(R) = -\frac{\mu_1}{R}, \qquad (4.9)$$

where $R \geq R_1$ is the distance from the center of the primary. This spherically symmetric potential would produce an ocean of uniform depth d over the entire spherical planet. However, in the presence of a secondary, we must add the tidal potential from Equation 4.8:

$$\Phi(R, \psi) \approx -\frac{\mu_1}{R} - \frac{\mu_2}{a} \left(\frac{R}{a'} \right)^2 P_2(\cos \psi). \qquad (4.10)$$

The addition of the tidal potential means that the equipotential surfaces are not spherical, and therefore the depth of the ocean is not uniform. If the depth in the absence of the secondary is $d \ll R_1$, the depth in the presence of the secondary is $d + \delta R(\psi)$. (We have been careful to pour a deep enough ocean to ensure that $d + \delta R$ is positive everywhere, and the ocean floor is never exposed.)

To find $\delta R(\psi)$, we start with Equation 4.10 for the potential, then make the substitution

$$R = R_0 \left[1 + \frac{\delta R}{R_0} \right], \qquad (4.11)$$

where $R_0 \equiv R_1 + d$ is the mean distance of the ocean surface from the center of the primary. With this substitution, and the expectation that $|\delta R/R_0| \ll 1$, we can write Equation 4.10 as

$$\Phi(\delta R, \psi) \approx -\frac{\mu_1}{R_0} + \frac{\mu_1}{R_0} \left(\frac{\delta R}{R_0} \right) - \frac{\mu_2}{a} \left(\frac{R_0}{a} \right)^2 P_2(\cos \psi). \qquad (4.12)$$

In Equation 4.12, the first term on the right-hand side represents the primary's gravitational potential at the mean ocean surface. The second term represents the (small) change in the primary's gravitational potential as the ocean surface rises or falls relative to the mean surface. The third term represents the (small) tidal potential from the secondary. To locate an equipotential surface, we must find the value of $\delta R(\psi)$ for which the second and third terms cancel. That is, we must solve the equation

$$\frac{\mu_1}{R_0}\left(\frac{\delta R}{R_0}\right) = \frac{\mu_2}{a}\left(\frac{R_0}{a}\right)^2 P_2(\cos\psi) \tag{4.13}$$

for $\delta R(\psi)$. We can write the solution conveniently as

$$\frac{\delta R}{R_1} = \varepsilon_T P_2(\cos\psi), \tag{4.14}$$

where the dimensionless tidal amplitude ε_T is

$$\varepsilon_T = \frac{m_2}{m_1}\left(\frac{R_0}{a}\right)^3 \approx \frac{m_2}{m_1}\left(\frac{R_1}{a}\right)^3. \tag{4.15}$$

(Since we have added only a shallow ocean, the difference between the primary's mean radius R_0 with the ocean and its radius R_1 without the ocean is negligible.) The Legendre polynomial $P_2(\cos\psi)$ has maxima at $\psi = 0$ and $\psi = \pi$, where $P_2 = 1$; its minimum is at $\psi = \pi/2$, where $P_2 = -1/2$. The ocean is thus stretched into a prolate spheroid with its major axis pointing toward and away from the secondary. The difference in height between the maximum and minimum values of δR is then

$$h_T = \frac{3}{2}\varepsilon_T R_1 = \frac{3}{2}\frac{m_2}{m_1}\left(\frac{R_1}{a}\right)^3 R_1. \tag{4.16}$$

In an ideal ocean world, the length scale h_T is the difference in the height of the ocean surface between low tide and high tide. Although the Earth is not an ideal ocean world, we can can still compute h_T and see whether it roughly corresponds to observed reality. For the Earth–Moon system, $m_2/m_1 = 0.0123$ and $R_1 = 6370\,\text{km} = 0.0166a$. This yields $\varepsilon_T = 5.6 \times 10^{-8}$ for the dimensionless strength of the tide, and $h_T = 0.54\,\text{m}$ for its physical height. Although half a meter is the right order of magnitude for the height of tides in the open ocean, there are obvious complicating factors. One is the presence of continents obstructing the flow of water. Nearly landlocked seas, like the Mediterranean, have tidal variations of low amplitude ($h \sim 0.1\,\text{m}$). Conversely, long but shallow bays, like the Bay of Fundy, can display resonant behavior that creates tidal variations of high amplitude ($h \sim 10\,\text{m}$).

Another complication for the Earth's tides is the existence of the Sun. Although the Sun is much further from us than the Moon is, it is also very much more massive than the Moon. Let's compute the height of the Earth's tidal bulges as raised by the Sun. In the Sun–Earth system, the Earth is the secondary. The relevant mass ratio is $m_1/m_2 = 3.33 \times 10^5$, and the Earth's radius is $R_2 = 6370\,\text{km} = 4.26 \times 10^{-5}\,\text{au}$. This yields $\varepsilon_T = 2.6 \times 10^{-8}$ and

$h_T = 0.25\,\text{m}$ for the height of the Earth's tidal bulges due to the Sun, about 46% of the height of the tidal bulges raised by the Moon. As early cultures realized, the "spring tides" that occur during full Moon and new Moon, when the tidal bulges raised by the Sun and Moon align, are higher in amplitude than the "neap tides" that occur during first or last quarter phase.

Another complication in the Earth's tides is that the so-called *terra firma* beneath our feet is not a rigid solid. The Earth's crust is fairly rigid, but it lies on a partially fluid mantle. Thus, even when you are on solid ground, you slowly bob up and down through a total height $h \sim 0.3\,\text{m}$ over the course of 12 hours, 25 minutes. Thus, even objects without a liquid ocean are subject to tidal distortion, with consequences that we examine in the next section.

One final aside: Since gases as well as liquids are fluids, gaseous atmospheres also tend to stratify themselves on equipotential surfaces. However, for most planets with an atmosphere, the expansion and contraction of the atmosphere during the day/night cycle is much larger in amplitude than the tidal height h_T.

4.2 Tidal Heating

The height of the Earth's tidal bulges is $\sim 10^{-7}$ times the Earth's mean radius. You might wonder why we care about such a tiny effect. The Earth's tidal bulges, after all, are dwarfed by its equatorial bulge, which has a height $\sim 3 \times 10^{-3}$ times the Earth's mean radius. We could start by pointing at other objects that have larger tidal distortions. Since gravity is a mutual force, the Earth creates tidal bulges on the Moon. To compute the height of tides on the Moon, we take $m_1/m_2 = 81.3$ for the mass ratio, and $R_2 = 1740\,\text{km} = 4.52 \times 10^{-3}a$ as the radius of the tidally disturbed body. These numbers yield $\varepsilon_T = 7.5 \times 10^{-6}$ for the dimensionless strength of tides on the Moon, and $h_T = 20\,\text{m}$ as the height of the Moon's tides (if the outer layer of the Moon were a perfect fluid).

Even within the solar system, there are two-body systems that show stronger tidal effects than the Earth–Moon system. Table 4.1 shows the values of ε_T and h_T for two of these tidally affected pairings: the Jupiter–Io system and the Mars–Phobos system. Although Io is about the same size as the Moon, and the Jupiter–Io distance is comparable to the Earth–Moon distance, the great mass of Jupiter (over 300 times the Earth's mass) means that Io has tidal bulges with $h_{T,2} \approx 4700\,\text{m}$. Although the planet Mars is a lightweight compared to Jupiter, its inner moon Phobos is at a distance of only 2.77 Martian radii from the center of Mars. This means that Phobos has $\varepsilon_{T,2} \sim 0.1$, giving it huge tidal bulges compared to its size.[3] Note that if the primary and secondary in

---

[3] The tidal effects of Mars on Phobos are complicated by the fact that Phobos is intrinsically non-spherical, with axis ratios $1:0.88:0.70$; the values in Table 4.1 are approximate, and assume an intrinsically spherical Phobos.

Table 4.1 Parameters for selected two-body systems.

	Units	Earth–Moon	Jupiter–Io	Mars–Phobos
$\varepsilon_{T,1}$		5.602×10^{-8}	2.159×10^{-7}	7.848×10^{-10}
$\varepsilon_{T,2}$		7.503×10^{-6}	1.713×10^{-3}	~ 0.10
$h_{T,1}$	m	0.5354	22.65	3.990×10^{-3}
$h_{T,2}$	m	19.55	4680	~ 180
ω_1	$10^{-5}\,\mathrm{s}^{-1}$	7.292	17.59	7.088
ω_2	$10^{-5}\,\mathrm{s}^{-1}$	0.2662	4.111	22.80
n	$10^{-5}\,\mathrm{s}^{-1}$	0.2662	4.111	22.80
$E_{\mathrm{rot},1}$	$10^{36}\,\mathrm{erg}$	2.133	3.643×10^5	6.749×10^{-2}
$E_{\mathrm{rot},2}$	$10^{36}\,\mathrm{erg}$	3.087×10^{-6}	9.471×10^{-4}	$\sim 1.8 \times 10^{-13}$
$|E_{\mathrm{orb}}|$	$10^{36}\,\mathrm{erg}$	0.3810	1.342×10^2	2.434×10^{-7}
$J_{\mathrm{rot},1}$	$10^{40}\,\mathrm{g\,cm}^2\,\mathrm{s}^{-1}$	5.851	4.144×10^5	0.1904
$J_{\mathrm{rot},2}$	$10^{40}\,\mathrm{g\,cm}^2\,\mathrm{s}^{-1}$	2.320×10^{-4}	4.608×10^{-3}	$\sim 1.6 \times 10^{-13}$
J_{orb}	$10^{40}\,\mathrm{g\,cm}^2\,\mathrm{s}^{-1}$	28.37	6.528×10^2	2.316×10^{-7}

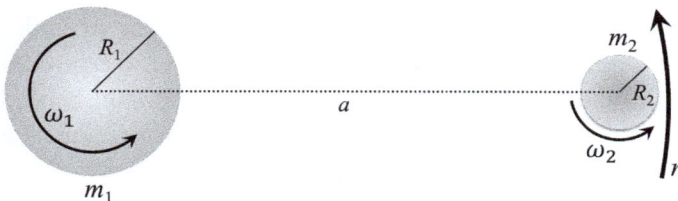

Figure 4.2 A primary has rotational angular speed ω_1, while a secondary has rotational angular speed ω_2. The orbital mean motion of the system is n. All angular momentum vectors are assumed to be parallel.

a system are of comparable density, then Equations 4.15 and 4.16 tell us that $\varepsilon_{T,1}/\varepsilon_{T,2} \sim m_2/m_1$ and $h_{T,1}/h_{T,2} \sim (m_2/m_1)^{2/3}$. Thus, for the Mars–Phobos system, with its large mass ratio, the 100 m tidal bulges on Phobos dwarf the 4 mm bulges on Mars.

Even when tidal bulges are small, they can have a significant effect because they are *dynamic*, in comparison to static equatorial bulges or intrinsic triaxiality. Let's consider the simple two-body system shown in Figure 4.2. The nearly spherical primary is rotating with angular speed ω_1.[4] The secondary is also nearly spherical, and rotates with angular speed ω_2. For simplicity, we assume that all angular momentum vectors in the problem are parallel; that is, the orbital plane is the same as the equatorial planes of the primary and secondary.

---

[4] We use a lower-case omega (ω) for rotational angular speed, to distinguish it from the upper-case Omega (Ω) used in Chapter 2 for orbital angular speed.

First, consider the primary. If $\omega_1 \neq n$, then an observer on the primary's surface experiences **semidiurnal tides**. The time between high tides is

$$\mathcal{P}_T = \frac{\pi}{|\omega_1 - n|}. \tag{4.17}$$

For tides raised on the Earth by the Moon, the values of ω_1 and n listed in Table 4.1 lead to $\mathcal{P}_T = 4.471 \times 10^4 \text{s} = 12.42 \text{hr}$. As the observer on the primary's surface bobs up and down with the tides, the motions within the primary cause heating due to internal friction. On the Earth, most of this friction occurs in the oceans, with smaller contributions from the earth below and atmosphere above. The Earth's heating rate from tidal friction is $L_T = 3.7 \times 10^{19} \text{erg s}^{-1}$. Although the Earth's heating rate from radioactive decay is about five times this value, the tidal heating is not completely negligible.

In many close two-body systems, it is found that the secondary has a rotational angular speed ω_2 that is equal to the mean motion n. This is true, as Table 4.1 shows, for Io and Phobos as well as for our own familiar Moon. When $\omega = n$, an object is referred to as having **synchronous rotation**. In some close binaries, such as the Pluto–Charon system, both the secondary and primary are in synchronous rotation, with $\omega_1 = \omega_2 = n$. Despite their lack of semidiurnal tides, objects in synchronous rotation can still be tidally heated if they are on eccentric orbits. The Earth's Moon, for instance, has an orbit with $e = 0.055$. Thus, its tidal bulges are bigger at periapsis than at apoapsis by an amount

$$\frac{h_{pe}}{h_{ap}} = \left(\frac{r_{ap}}{r_{pe}}\right)^3 = \left(\frac{1+e}{1-e}\right)^3 = 1.39. \tag{4.18}$$

Instead of having a constant $h_T = 19.55 \text{m}$, as computed above, the Moon's tidal bulges vary in height from $h_{pe} \approx 23.2 \text{m}$ at periapsis to $h_{ap} \approx 16.6 \text{m}$ at apoapsis. The stretching and squeezing of the tidal bulges over the course of a sidereal month causes tidal heating by internal friction.

The tidal heating of the Moon is tiny compared to that of Io. The Laplace resonance among Io, Europa, and Ganymede forces Io's orbit to be slightly eccentric, with $e \approx 0.004$. Thus, Io's tidal bulges, instead of having a constant $h_T = 4680 \text{m}$, vary in height from $h_{pe} = 4740 \text{m}$ at periapsis to $h_{ap} = 4620 \text{m}$ at apoapsis. The resulting stretch and squeeze through $\Delta h \approx 120 \text{m}$ (compared to $\Delta h \approx 6.6 \text{m}$ for the Moon) causes significant tidal heating. Observations of volcanic activity on Io yield an estimate of $L \approx 10^{21} \text{erg s}^{-1}$ for the heat flux through Io's surface. To replace this energy lost to space in the form of infrared photons, tidal heating must provide $L_T \approx 10^{21} \text{erg s}^{-1}$, nearly 30 times the Earth's tidal heating rate.

In a close binary system, one or both objects can have significant tidal heating. This tidal heating will result from semidiurnal tides if $\omega \neq n$, or from orbital eccentricity if $e > 0$. In the extreme case of Io, maintaining tidal heating for the age of the solar system ($t \approx 4.6\,\mathrm{Gyr} \sim 10^{17}\,\mathrm{s}$) requires an energy input $E_T \sim L_T t \sim 10^{38}\,\mathrm{erg}$. Where does the energy for tidal heating ultimately come from? One possible energy reservoir is the orbital energy E_{orb} of the system. The sum of the gravitational potential energy and the orbital kinetic energy, given in Equation 1.54, is

$$E_{\mathrm{orb}} = -\frac{Gm_1 m_2}{2a}. \tag{4.19}$$

For the Jupiter–Io system, this energy is $E_{\mathrm{orb}} \approx -1.3 \times 10^{38}\,\mathrm{erg}$. From a purely energetic level, the target energy $E_T \sim 10^{38}\,\mathrm{erg}$ is similar to the energy released by letting Io spiral inward from a much larger orbit. However, we still need to examine another possible energy source: the rotational kinetic energy of the two objects in the system.

Let's approximate each object in a two-body system as a sphere of mass m and radius R in solid body rotation with angular speed ω. A sphere's moment of inertia around its rotation axis can be written as

$$I_{\mathrm{rot}} = K_I m R^2, \tag{4.20}$$

where the dimensionless constant K_I is a measure of how centrally concentrated the mass is. For a sphere of uniform density, $K_I = 0.4$. Terrestrial planets and rocky moons usually lie in the range $K_I = 0.33$–0.39. Gas giants like Jupiter and Saturn, and ice giants like Uranus and Neptune, are more centrally concentrated, and have $K_I = 0.22$–0.28. Gaseous stars are even more concentrated, and have $K_I \approx 0.07$ for a main sequence star like the Sun. The rotational kinetic energy of a sphere can then be written as

$$E_{\mathrm{rot}} = \frac{1}{2} I_{\mathrm{rot}} \omega^2 = \frac{K_I}{2} m R^2 \omega^2. \tag{4.21}$$

The rotational kinetic energy $E_{\mathrm{rot},2}$ of the secondary in a binary system is usually small. If we use Equation 4.21 to compute the secondary's rotational kinetic energy, then divide by its orbital energy (Equation 4.19), we find

$$\frac{E_{\mathrm{rot},2}}{|E_{\mathrm{orb}}|} \approx K_I \left(\frac{R_2}{a}\right)^2 \left(\frac{\omega_2}{n}\right)^2. \tag{4.22}$$

For secondaries in synchronous rotation, such as Io and the Moon, we set $\omega_2 = n$ in Equation 4.22, leaving

$$\frac{E_{\mathrm{rot},2}}{|E_{\mathrm{orb}}|} \approx K_I \left(\frac{R_2}{a}\right)^2 \ll 1. \tag{4.23}$$

The primary in a close binary, however, is often rapidly rotating, with $\omega_1 > n$. This rapid rotation, combined with the large mass of the primary, makes it an effective flywheel for the storage of rotational kinetic energy. In the Jupiter–Io system, for instance, Jupiter has $\omega_1 \approx 4.3n$. Its mass is 21 000 times that of Io, while its radius is 38 times that of Io. Putting it all together, Jupiter's rotational kinetic energy is

$$E_{\text{rot},1} \approx 3.6 \times 10^{41} \text{ erg} \approx 2700|E_{\text{orb}}| \approx 3.8 \times 10^8 E_{\text{rot},2}. \tag{4.24}$$

The Jupiter–Io system displays an energy hierarchy that is common in close binaries with $f \ll 1/2$; the rotational kinetic energy of the primary is larger than the magnitude of the orbital energy, which in turn is larger than the rotational kinetic energy of the small, low-mass secondary. Table 4.1, for instance, shows that this hierarchy holds true for the Earth–Moon system and for the Mars–Phobos system.

4.3 Tides and Orbital Evolution

Tidal heating can measurably change the orbital and rotational properties of a two-body system. Consider the system shown in Figure 4.2, in which a nearly spherical primary of mass m_1 and radius R_1 rotates with angular speed ω_1. The rotational angular momentum of the primary is

$$J_{\text{rot},1} = I_{\text{rot},1}\omega_1, \tag{4.25}$$

where $I_{\text{rot},1} = K_1 m_1 R_1^2$. The rotational kinetic energy of the primary can be written as

$$E_{\text{rot},1} = \frac{1}{2}I_{\text{rot},1}\omega_1^2 = \frac{1}{2}J_{\text{rot},1}\omega_1. \tag{4.26}$$

Similar equations, with a change in subscript, give the rotational angular momentum $J_{\text{rot},2}$ and rotational kinetic energy $E_{\text{rot},2}$ for the secondary. We assume the secondary is on a circular orbit in the equatorial plane of the primary. In this case, the orbital angular momentum of the system, with magnitude

$$J_{\text{orb}} = \frac{m_1 m_2}{m_1 + m_2}\mu^{1/2}a^{1/2} \approx m_2\mu^{1/2}a^{1/2}, \tag{4.27}$$

is parallel to the rotational angular momentum J_{rot} of the primary. Real-universe examples of $J_{\text{rot},1}$, $J_{\text{rot},2}$, and J_{orb} are given in Table 4.1 for the Earth–Moon, Jupiter–Io, and Mars–Phobos systems. Note that for these systems, $J_{\text{rot},2}$ is tiny compared to both $J_{\text{rot},1}$ and J_{orb}.

The total energy pool available for conversion to heat, and then to radiated photons, is

$$E = E_{\text{rot},1} + E_{\text{orb}} = \frac{1}{2}I_{\text{rot},1}\omega_1^2 - \frac{Gm_1 m_2}{2a}. \tag{4.28}$$

We ignore the tiny rotational kinetic energy of the secondary, and assume that the primary and secondary are in thermal equilibrium, with internal thermal energy staying in a balance between tidal heating and radiative cooling. If m_1, m_2, and $I_{\text{rot},1}$ are constant, the rate of energy loss from the system is

$$\frac{dE}{dt} = I_{\text{rot},1}\omega_1 \frac{d\omega_1}{dt} + \frac{Gm_1m_2}{2a^2}\frac{da}{dt} < 0. \tag{4.29}$$

Thus, losses from tidal heating must be accompanied by slowing of the primary's rotation or by shrinking of the secondary's orbit.

The link between the changes in ω_1 and a is provided by the fact that the total angular momentum J of the system is conserved. Assuming that the rotational angular momentum of the secondary is negligible, the total angular momentum is

$$J = J_{\text{rot},1} + J_{\text{orb}} = I_{\text{rot},1}\omega_1 + m_2\mu^{1/2}a^{1/2}. \tag{4.30}$$

Conservation of angular momentum requires that $dJ/dt = 0$ and thus

$$I_{\text{rot},1}\frac{d\omega_1}{dt} = -m_2\frac{\mu^{1/2}}{2a^{1/2}}\frac{da}{dt}. \tag{4.31}$$

If the rotation of the primary slows, the orbit of the secondary must expand to conserve angular momentum; conversely, if the rotation speeds up, the orbit of the secondary must shrink. Substituting from Equation 4.31 into Equation 4.29, we find a relation among energy loss, rotational speed, and orbital size:

$$\frac{dE}{dt} = \left(\omega_1 - \frac{\mu^{1/2}}{a^{3/2}}\right)I_{\text{rot},1}\frac{d\omega_1}{dt} = (\omega_1 - n)I_{\text{rot},1}\frac{d\omega_1}{dt}. \tag{4.32}$$

Consider the Earth–Moon system, in which $\omega_1 - n > 0$; the Earth rotates once per sidereal day, while the laggard Moon orbits once per sidereal month. In this case, the loss of energy ($dE/dt < 0$) requires the Earth's rotation to slow ($d\omega_1/dt < 0$) and the Moon to drift gradually away ($da/dt > 0$). The rate at which the Earth's rotation is slowing is measurable; from the observed times of eclipses over the past three millennia, the rate of slowing is

$$\frac{d\omega_1}{dt} = -(4.77 \pm 0.08) \times 10^{-22}\,\text{s}^{-2}. \tag{4.33}$$

Given the Earth's current rotation speed $\omega_1 = 7.29 \times 10^{-5}\,\text{s}^{-1}$, this yields a tidal braking timescale of

$$\left|\frac{1}{\omega_1}\frac{d\omega_1}{dt}\right|^{-1} = (1.53 \pm 0.03) \times 10^{17}\,\text{s} = 4.84 \pm 0.08\,\text{Gyr}. \tag{4.34}$$

Conservation of angular momentum, in the form of Equation 4.31, requires that the Moon's orbit expand at the rate

$$\frac{da}{dt} = -\frac{2}{m_2}\frac{a^{1/2}}{\mu^{1/2}}I_{rot,1}\frac{d\omega_1}{dt} \tag{4.35}$$

$$= -\frac{2}{m_2}\frac{1}{na}I_{rot,1}\frac{d\omega_1}{dt} = 3.16 \pm 0.05\,\text{cm}\,\text{yr}^{-1},$$

given a moment of inertia $I_{rot} = 8.024 \times 10^{44}\,\text{g}\,\text{cm}^2$ for the Earth. The Lunar Laser Ranging experiment, using reflectors left on the Moon by the *Apollo* astronauts, measured the rate of expansion for the Moon's orbit as

$$\frac{da}{dt} = 3.82 \pm 0.07\,\text{cm}\,\text{yr}^{-1}. \tag{4.36}$$

The statistically highly significant difference between the predicted rate in Equation 4.35 and the measured rate in Equation 4.36 can be attributed to a breakdown in our assumption that $I_{rot,1}$ is constant. Earthquakes tend to "compactify" the Earth, reducing its moment of inertia; the melting of icecaps also changes the Earth's moment of inertia.

The slowing of the Earth's rotation and the receding of the Moon will continue until $\omega_1 = n$, and the Earth–Moon system attains a stable equilibrium. This will occur when $\mathcal{P}_{rot,1} = \mathcal{P}_{rot,2} = \mathcal{P}_{orb} \approx 47.4\,\text{d}$. When this equilibrium is reached, the Moon will be $\sim$44% farther away than it is now, and the orbital angular momentum will account for more than 99.6% of the total angular momentum of the Earth–Moon system.[5]

As a contrast to the Earth–Moon system, consider the Mars-Phobos system, where the orbital period of Phobos ($P = 7.65\,\text{hr}$) is much shorter than the rotation period of Mars ($\mathcal{P}_{rot,1} = 24.62\,\text{hr}$). In this case, where $\omega_1 - n < 0$, Equation 4.32 tells us that the loss of energy from thermal radiation ($dE/dt < 0$) requires the rotation of Mars to speed up ($d\omega_1/dt > 0$) and Phobos to drift gradually inward ($da/dt < 0$). Phobos has orbited Mars $\sim$200 000 times since it was discovered in 1877; given this long baseline for observation, its orbital parameters are quite well determined. In particular, the mean motion of Phobos' orbit is currently increasing at the rate

$$\frac{dn}{dt} = (4.451 \pm 0.011) \times 10^{-20}\,\text{s}^{-2}. \tag{4.37}$$

From Kepler's third law, this implies that the semimajor axis is shrinking at the rate

$$\frac{da}{dt} = -\frac{2}{3}\frac{a}{n}\frac{dn}{dt} = -3.85 \pm 0.01\,\text{cm}\,\text{yr}^{-1}. \tag{4.38}$$

---

[5] We have already computed the timescale for this to happen as $\sim$5 Gyr, so don't go planning your observing strategies for three-week-long nights just yet.

Since the periapsis of Phobos' orbit is currently at a height $d \approx 5800$ km above the Martian surface, this would argue for a timescale $t \sim 150$ Myr before Phobos plows into Mars. However, it is likely that the tidal forces on Phobos will tear it apart long before then, creating a ring around Mars.

4.4 Tidal Torque

In the previous section, we examined how the change of a primary's rotation was related to the change of the secondary's orbit. We did this purely by relying on conservation of energy and angular momentum, averting our eyes from the details of how the necessary torques were applied. In some circumstances, relying purely on conservation laws is justifiable; you don't need to know a pool ball's coefficient of restitution to put the eight ball in the corner pocket. However, in the case of tidal evolution in a binary system, it is useful to know some of the physics behind **tidal torques**.

Start by considering a system in which the primary is our idealized ocean world. The secondary is on a circular orbit in the equatorial plane of the primary. If the oceans of the primary are made of a perfect, viscosity-free, frictionless liquid, then an observer on the equator will experience high tide at the moments when the secondary is at the zenith, directly overhead, and at the nadir, directly underfoot. However, real planets have friction, as do real moons and real stars. Figure 4.3(a) shows a system in which friction exists, and for which $\omega_1 - n > 0$; this could be the Earth–Moon system, for instance. Given internal friction within the primary, the tidal bulge closest to the secondary slightly leads the secondary. For the Earth–Moon system, the misalignment angle is $\epsilon_{\mathrm{mis}} \approx 3° \approx 0.052$ rad. Thus, an observer on

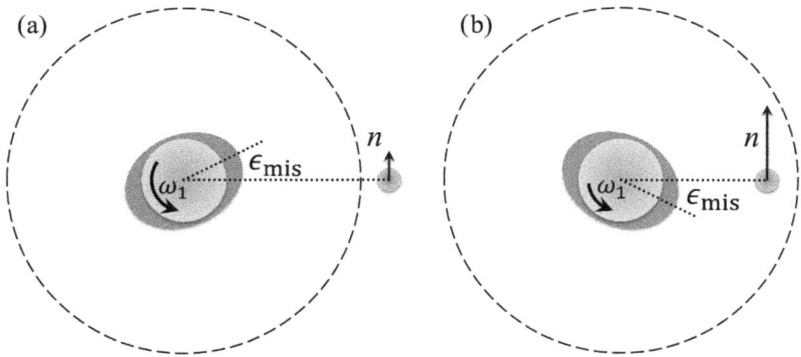

Figure 4.3 (a) Planet–moon system with $\omega_1 > n$; the tidal bulge closest to the moon leads the moon on its orbit. (b) Planet–moon system with $\omega_1 < n$; the closest tidal bulge trails the moon. (The height of the tidal bulges is wildly exaggerated for clarity.)

Earth experiences high tide when the Moon is ~3° west of the meridian (or about 12 minutes after meridian passage).[6] Figure 4.3(b) shows a system with $\omega_1 - n < 0$; this could represent the Mars–Phobos system. In this case, the bulge lags behind the secondary.

The lack of perfect alignment is, like tidal heating, a result of internal friction within the primary. Let's look once more at the energy budget associated with tides. If the Earth were an ideal ocean world, the gravitational potential energy stored in its tidal bulges would be comparable to the energy required to take a layer of the ocean $\sim h_T/2$ deep, covering half the planet, and lift it by a distance $\sim h_T/2$. This gives an order-of-magnitude result

$$E_{\text{bulge}} \approx \left(2\pi R_1^2 \frac{h_T}{2}\rho\right)\left(g\frac{h_T}{2}\right) \approx \frac{\pi}{2}\mu_1\rho h_T^2, \tag{4.39}$$

where ρ is the density of the ocean and $g = \mu_1/R_1^2$ is the gravitational acceleration at the ocean surface. Using the Earth's mass, and assuming an ocean with $\rho = 1\,\text{g cm}^{-3}$ and $h_T = 54\,\text{cm}$, the energy scale is

$$E_{\text{bulge}} \sim 2 \times 10^{24}\,\text{erg}. \tag{4.40}$$

This is equivalent to ~50 megatons of TNT; in some contexts, this is a lot of energy, but not in the context of Earth's tidal heating. The amount of tidal heat radiated by the Earth during one cycle of tides, given $dE/dt = 3.7 \times 10^{19}\,\text{erg s}^{-1}$, is

$$\Delta E = \frac{dE}{dt}\mathcal{P}_T = 1.7 \times 10^{24}\,\text{erg}. \tag{4.41}$$

The effectiveness with which tidal heating dissipates energy is conventionally given by the **tidal quality factor**, defined as

$$Q \equiv \frac{2\pi E_{\text{bulge}}}{\Delta E}, \tag{4.42}$$

where E_{bulge} is the energy associated with tidal distortions (such as the Earth's tidal bulges), and ΔE is the energy radiated away during one tidal cycle. Using the value of E_{bulge} estimated above, we expect the Earth's tidal quality factor to be

$$Q \sim \frac{2\pi(2 \times 10^{24}\,\text{erg})}{1.7 \times 10^{24}\,\text{erg}} \sim 7. \tag{4.43}$$

A frictionless ocean world, by contrast, would have $dE/dt = 0$ and thus $Q = \infty$.

Tidal dissipation generally occurs by complicated viscoelastic forces within celestial objects, and depends on the frequency of the tidal perturbation. The value of Q is usually deduced indirectly from astronomical observations. For

---

[6] On your next vacation by the sea, you are likely to experience a time delay greater than or less than 12 minutes; this is another side effect of having continents.

instance, the misalignment angle ϵ_{mis} in Figure 4.3 is inversely related to Q. In the limit $Q \gg 1$, the relation between the misalignment and Q is

$$\epsilon_{mis} = \frac{1}{2Q}. \tag{4.44}$$

Thus, the Earth's misalignment angle relative to the Moon leads us to deduce that $Q = 0.5/\epsilon_{mis} \approx 10$ at a tidal period $\mathcal{P}_T = 12.42\,\mathrm{hr}$; this is a more reliable value for Q than our hand-waving estimate above. The value of Q for the Earth is small compared to that of a rocky object without oceans. The Moon, for instance, undergoes a complete tidal cycle once per sidereal month ($\mathcal{P}_T = 27.32\,\mathrm{d}$) because of the eccentricity of its orbit; its tidal quality parameter at this period is $Q \approx 38$. Mars, by comparison, has $Q \approx 90$ at the period $\mathcal{P}_T = 5.55\,\mathrm{hr}$ of the tides raised by Phobos.

Consider a system in which the tidal bulges of the primary are misaligned with the primary–secondary axis by a small angle $\epsilon_{mis} = 1/(2Q)$, as shown in Figure 4.3. We can make an approximate calculation of the primary's quadrupole moment by assuming it is a prolate spheroid of uniform density; its long semi-axis is $A = R_1 + 2h_T/3$ and its short semi-axes are $B = C = R_1 - h_T/3$. Using the standard formula for the quadrupole moment of a uniform density prolate spheroid,

$$q_1 = \frac{2}{5}m_1(A^2 - C^2), \tag{4.45}$$

we find in the small-bulge limit ($h_T \ll R_1$),

$$q_1 \approx \frac{4}{5}m_1 R_1 h_T \approx \frac{6}{5}m_2 R_1^2 \left(\frac{R_1}{a}\right)^3 \qquad \text{[ideal]}, \tag{4.46}$$

using Equation 4.16 for the tidal height h_T.

However, real primaries are not uniform in density, and real tidal bulges are not equal in height to h_T. In general, the quadrupole moment of real tidally distorted bodies can be written as

$$q_1 = 2k_T m_2 R_1^2 \left(\frac{R_1}{a}\right)^3 \qquad \text{[real]}, \tag{4.47}$$

where the dimensionless parameter k_T is the **tidal Love number**.[7] For the Earth, the tidal Love number is $k_T = 0.295$, for example. For gas giants and ice giants, k_T usually lies in the range $k_T = 0.1$–0.6. For Sun-like stars, the strong central concentration of mass yields a small tidal Love number, $k_T \approx 0.03$.

---

[7] The tidal Love number k_T is often written as k_2, since it encodes information about the $\ell = 2$ mass moment. It is named after the mathematician A. E. H. Love; thus "tidal Love" is not a reference to your expected emotional state when studying tides.

The gravitational potential near the tidally distorted primary, given its non-zero quadrupole moment q_1, is

$$\Phi_1(R, \psi) = -\frac{Gm_1}{R} - \frac{1}{2}\frac{Gq_1}{R^3}P_2\left(\cos(\psi - \epsilon_{\mathrm{mis}})\right),\qquad(4.48)$$

where $R \geq R_1$ is the distance from the primary's center and the angle ψ is measured relative to the line connecting the primary's center to the secondary's center. The quadrupole component of the potential, making use of Equation 4.47, can be written as

$$\Phi_{\mathrm{quad},1}(R, \psi) = -k_{\mathrm{T}}\frac{\mu_2}{R^3}R_1^2\left(\frac{R_1}{a}\right)^3 P_2\left(\cos(\psi - \epsilon_{\mathrm{mis}})\right).\qquad(4.49)$$

On the surface of the primary, where $R = R_1$, the quadrupole component of the primary's potential can be written as

$$\Phi_{\mathrm{quad},1}(R_1, \psi) \approx -k_{\mathrm{T}}\frac{\mu_2}{a}\left(\frac{R_1}{a}\right)^2 P_2\left(\cos(\psi - \epsilon_{\mathrm{mis}})\right).\qquad(4.50)$$

However, from Equation 4.12, we can write the quadrupole component of the *secondary's* potential as

$$\Phi_{\mathrm{quad},2}(R_1, \psi) \approx -\frac{\mu_2}{a}\left(\frac{R_1}{a}\right)^2 P_2(\cos\psi).\qquad(4.51)$$

Thus, at the surface of the tidally distorted primary, there are actually two quadrupole potentials, shifted in phase by ϵ_{mis}. The potential $\Phi_{\mathrm{quad},2}$ is due to the presence of a secondary at a distance a from the primary; the potential $\Phi_{\mathrm{quad},1}$ is due to the resulting tidal distortion of the primary.[8] The ratio of the magnitude of the two quadrupole moments from Equations 4.50 and 4.51 is

$$\frac{|\Phi_{\mathrm{quad},1}|}{|\Phi_{\mathrm{quad}\,2}|} = k_{\mathrm{T}}.\qquad(4.52)$$

This ratio is how Love originally defined the tidal Love number, and explains the seemingly arbitrary factor of two in Equation 4.47.

Since the misalignment angle ϵ_{mis} is non-zero, the quadrupole moment of the distorted primary exerts a torque on the secondary. The value of the tidal torque is

$$\frac{dJ_{\mathrm{orb}}}{dt} = -m_2\frac{\partial\Phi_{\mathrm{quad},1}}{\partial\psi}\bigg|_{R=a,\psi=0}.\qquad(4.53)$$

(The torque on the primary from the quadrupole moment of the secondary is equal in magnitude but opposite in sign: $dJ_{\mathrm{rot},1}/dt = -dJ_{\mathrm{orb}}/dt$.) Using the

---

[8] We ignored $\Phi_{\mathrm{quad},1}$ when examining an ideal ocean world in Section 4.1. This is because we assumed the mass of the tidally distorted ocean was negligible, leading to $k_{\mathrm{T}} \ll 1$.

primary's quadrupole potential from Equation 4.49, we find that in the limit $\epsilon_{mis} \ll 1$,

$$\left. \frac{\partial \Phi_{quad,1}}{\partial \psi} \right|_{\psi=0} \approx 3\epsilon_{mis}\Phi_{quad,1}, \tag{4.54}$$

leading to a torque linearly dependent on the misalignment angle. Using the relation $\epsilon_{mis} = 1/(2Q)$, Equations 4.53 and 4.54 combine to yield a torque

$$\frac{dJ_{orb}}{dt} \approx m_2 \frac{3}{2} \frac{k_T \mu_2}{Q} \frac{\mu_2}{a} \left(\frac{R_1}{a} \right)^5. \tag{4.55}$$

Since $J_{orb} \approx m_2 \mu_1^{1/2} a^{1/2}$, we can write the tidal evolution timescale for the secondary to drift outward (or inward, if $\omega_1 < n$) as

$$t_{TE} \equiv \left| \frac{1}{a} \frac{da}{dt} \right|^{-1} = \frac{1}{6\pi} \frac{Q}{k_T} \frac{m_1}{m_2} \left(\frac{a}{R_1} \right)^5 \mathcal{P}, \tag{4.56}$$

where $\mathcal{P}$ is the orbital period of the secondary. Note that the factors Q/k_T, m_1/m_2, and a/R_1 are all greater than unity, and all contribute to the fact that the timescale for tidally altering a, or equivalently n, is very much greater than the orbital period $\mathcal{P}$.

In the Earth–Moon system, $da/dt = 3.82 \pm 0.07 \, \mathrm{cm \, yr^{-1}}$ can be measured directly, leading to a tidal evolution timescale of

$$t_{TE} = 10.06 \pm 0.18 \, \mathrm{Gyr} = (1.345 \pm 0.025) \times 10^{11} \mathcal{P}. \tag{4.57}$$

Using Equation 4.56, we can calculate

$$\frac{Q}{k_T} = 6\pi \frac{t_{TE}}{\mathcal{P}} \frac{m_2}{m_1} \left(\frac{R_1}{a} \right)^5 = 39.1 \pm 0.7. \tag{4.58}$$

For other binary systems, it is easier to measure dn/dt, or equivalently $d\mathcal{P}/dt$. The tidal evolution timescale is then

$$t_{TE} \equiv \left| \frac{1}{a} \frac{da}{dt} \right|^{-1} = \frac{3}{2} \left| \frac{1}{n} \frac{dn}{dt} \right|^{-1} = \frac{3}{2} \left| \frac{1}{\mathcal{P}} \frac{d\mathcal{P}}{dt} \right|^{-1}, \tag{4.59}$$

making use of Kepler's third law. For example, the Mars–Phobos system, using the value of dn/dt from Equation 4.37, has

$$t_{TE} = 243.5 \pm 0.6 \, \mathrm{Myr} = (2.789 \pm 0.007) \times 10^{11} \mathcal{P}, \tag{4.60}$$

leading to

$$\frac{Q}{k_T} = 539.2 \pm 1.3. \tag{4.61}$$

As we discuss more fully in Section 5.3, measurements of dn/dt for exoplanetary orbits can be used to compute Q/k_T for the star being orbited. These values of Q/k_T are found to be $\sim 10^5$ or more for hot Jupiters orbiting Sun-like stars. Stars are centrally concentrated, giving them a small tidal Love number k_T; they are also not very viscous, meaning their tidal heating is ineffective and their quality factor Q is large.

Exercises

4.1   Assume the Moon is a perfect sphere of radius $R_{moon} = 1737$ km and that the Sun is a perfect sphere with photospheric radius $R_\odot = 695\,700$ km. How long will it be until the angular size of the Moon, seen when the Moon is at perigee, is smaller than the angular size of the Sun, seen when the Earth is at aphelion? (This is when total solar eclipses will become a thing of the past.) Assume that the Earth's orbit remains unchanged ($a = 1$ au, $e = 0.0167$), and that the eccentricity of the Moon's orbit remains constant ($e = 0.0549$).

4.2   As shown in Table 4.1, the mean motion n of Io is smaller than the rotational angular speed ω_1 of Jupiter. Thus, Io is drifting outward as a result of its tidal interaction with Jupiter.
(a) What is the current semimajor axis a_{now} of Io's orbit?
(b) Suppose that when the solar system formed, Io was on an orbit around Jupiter for which n was infinitesimally smaller than ω_1. Assuming Jupiter's angular speed ω_1 has remained constant, what was the initial semimajor a_{init} of Io's orbit?
(c) For what value of Q/k_T would the time for Io to drift out from a_{init} to a_{now} equal the age of the solar system, $t = 4.57$ Gyr? This places a lower limit on the value of Q/k_T for Jupiter.
(d) If Io had started on an orbit for which n was infinitesimally *larger* than ω_1, how long would it take to drift inward to the cloud tops of Jupiter, given the value of Q/k_T calculated in part (c)?

4.3   Consider an idealized ocean world consisting of a perfect sphere of rigid rock, with radius R_1 and uniform density ρ_{rock}, covered with an ocean of viscosity-free liquid, with average depth $d \ll R_1$. The liquid of the ocean has a uniform (non-zero) density, with $\rho_{ocean} = f\rho_{rock}$, where $f < 1$. A secondary of mass m_1, at a distance a, tidally distorts the ocean.
(a) What is the mass quadrupole moment q_1 for this ocean world? You may assume that the height h_T of the tidal bulges is given by the ideal formula of Equation 4.16.
(b) Show that in the limit $f \ll 1$, the tidal Love number for this ocean world is $k_T \approx (3/5)f$.

5

Exoplanetary Systems

The revival of the heliocentric model by Copernicus in the sixteenth century led to speculation about planets orbiting other stars. In a heliocentric model, stars must show annual parallax as the Earth moves around the Sun. The fact that the parallactic angle p'' is too small to be detected by the unaided human eye ($p'' < 1\,\mathrm{arcmin}$) implied that stars are at a great distance ($d > 3000\,\mathrm{au}$). For the stars in the night sky to be visible at all, post-Copernican astronomers reasoned, they must be comparable in luminosity to the Sun. If stars are similar to the Sun in that respect, why shouldn't they be similar to the Sun in being orbited by planets?

In Giordano Bruno's dialog *De l'infinito universo e mondi*, the naïve student Elpino asks why, if the stars are Sun-like objects, we can't see the Earth-like planets orbiting them. The wise Philotheo, a stand-in for Bruno himself, has the answer: "We discern only the largest suns, immense bodies. But we do not discern the earths because, being much smaller, they are invisible to us." Even today, direct imaging of exoplanets (planets orbiting other stars) is extremely difficult. Thus, the most effective methods for detecting exoplanets are indirect. Some of these methods are based on Newtonian dynamics. For example, the **radial velocity** method looks for Doppler shifts in the light emitted by stars as they travel on their small exoplanet-induced orbits.

The **transit** method looks for the small dips in a star's apparent brightness as a conveniently opaque planet passes in front of it.

In this chapter, we will focus on the dynamics behind the discovery of exoplanets orbiting stars. Within the solar system, the label "planet" is applied to a range of objects, from the relatively small **terrestrial** planets (Mercury, Venus, Earth, and Mars) through the **ice giant** planets (Uranus and Neptune) to the relatively large **gas giant** planets (Jupiter and Saturn). When discussing the properties of exoplanets, we scale to the properties of the Earth when we expect a small exoplanet:

$$1\,R_\oplus = 6370\,\text{km} = 9.16 \times 10^{-3}\,R_\odot, \tag{5.1}$$

$$1\,M_\oplus = 5.97 \times 10^{27}\,\text{g} = 3.00 \times 10^{-6}\,M_\odot. \tag{5.2}$$

We scale to the properties of Jupiter when we expect a large exoplanet:

$$1\,R_{\text{jup}} = 11.0\,R_\oplus = 0.100\,R_\odot, \tag{5.3}$$

$$1\,M_{\text{jup}} = 318\,M_\oplus = 9.55 \times 10^{-4}\,M_\odot. \tag{5.4}$$

In contrast to a planet, a **star** is usually defined as a self-gravitating gas ball that is massive enough for fusion of hydrogen (^{1}H) to occur in its interior; theoretical models reveal that this requires a mass $m > 78\,M_{\text{jup}}$. Gas balls in the mass range $13\,M_{\text{jup}} < m < 78\,M_{\text{jup}}$ are able to fuse deuterium (^{2}H) but not ordinary hydrogen; objects in this range are called **brown dwarfs**. Gas balls with $m < 13\,M_{\text{jup}}$ are unable to fuse deuterium, and are usually classified as gas giant planets.

For our purposes, we can define a star–exoplanet system as a bound system in which the primary has $m_1 > 78\,M_{\text{jup}}$ and the secondary has $m_2 < 13\,M_{\text{jup}}$, resulting in a mass ratio $m_2/m_1 < 1/6$ and a mass fraction in the secondary $f = m_2/(m_1 + m_2) < 0.14$. However, many of the results in this chapter can also be applied to star–brown dwarf systems or brown dwarf–exoplanet systems.

5.1 Radial Velocity Detection

Consider a system that consists of a single star with mass m_1 and a single exoplanet with mass $m_2 < m_1/6$. Although many of the known planetary systems, including our own solar system, have multiple planets or multiple stars, we start with the simplest case. Even if we can't directly image the exoplanet in this simple two-body system, we can still hope to detect its gravitational effect on the star. Relative to the barycenter of the system, the star has an elliptical orbit with semimajor axis (Equation 1.60)

$$a_1 = fa = 748\,000\,\text{km} \left(\frac{f}{10^{-3}}\right)\left(\frac{a}{5\,\text{au}}\right). \tag{5.5}$$

Compared to the star's radius R_1, this orbital size is usually not very large, with

$$\frac{a_1}{R_1} = 1.08 \left(\frac{f}{10^{-3}}\right) \left(\frac{a}{5\,\text{au}}\right) \left(\frac{R_1}{1\,R_\odot}\right)^{-1}, \tag{5.6}$$

where $1\,R_\odot = 695\,700\,\text{km}$ is the Sun's radius. The orbital period of both the star and the exoplanet is

$$P = 2\pi \frac{a^{3/2}}{\mu^{1/2}} = 11.8\,\text{yr} \left(\frac{m_1 + m_2}{1\,M_\odot}\right)^{-1/2} \left(\frac{a}{5\,\text{au}}\right)^{3/2}, \tag{5.7}$$

where $\mu = G(m_1 + m_2)$.

For the moment, let's assume the orbits are circular. (Since some exoplanetary orbits are known to be highly eccentric, we'll relax this assumption later.) The star moves along its small circular orbit with a constant speed

$$v_1 = f\frac{\mu^{1/2}}{a^{1/2}} = 13.3\,\text{m s}^{-1} \left(\frac{f}{10^{-3}}\right) \left(\frac{m_1 + m_2}{1\,M_\odot}\right)^{1/2} \left(\frac{a}{5\,\text{au}}\right)^{-1/2}. \tag{5.8}$$

For any plausible star–exoplanet pair, this is a highly non-relativistic speed.

In general, as shown in Figure 5.1(a), an observer in the solar system sees the orbital plane of the star–exoplanet system inclined by some angle i relative to the plane of the sky. When the orbit is seen face-on, then $i = 0°$; when it is seen edge-on, then $i = 90°$.[1] For purposes of illustration, Figure 5.1 shows a case where $i = 60°$. An observer with high enough angular resolution can see the star move through a complete orbit during one orbital period $\mathcal{P}$. The projected orbit is an ellipse with angular semimajor axis

$$a_1'' = \frac{a_1}{d} = 5 \times 10^{-4}\,\text{arcsec} \left(\frac{f}{10^{-3}}\right) \left(\frac{a}{5\,\text{au}}\right) \left(\frac{d}{10\,\text{pc}}\right)^{-1}. \tag{5.9}$$

(a)   (b)

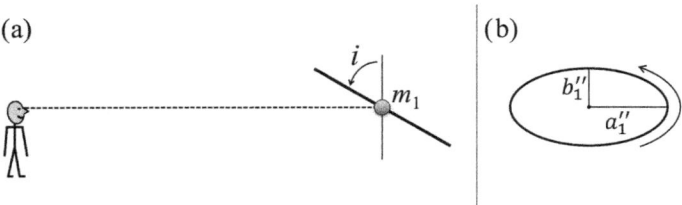

Figure 5.1 (a) The orbit of the star (heavy solid line) is inclined by an angle i relative to the plane of the sky (light solid line). (b) Stellar orbit as mapped by the observer; the barycenter of the star–exoplanet system is at the center of the ellipse.

[1] Sometimes inclinations $i > 90°$ are quoted. If the sense of orbital motion is known (clockwise or counterclockwise as seen by the observer), the adopted convention is that $0° \le i < 90°$ for counterclockwise motion and $90° < i \le 180°$ for clockwise motion.

The semiminor axis is then $b_1'' = a_1'' \cos i$. In addition, an observer on Earth sees the star move on a parallactic ellipse with angular semimajor axis

$$p'' = 0.1 \, \mathrm{arcsec} \left(\frac{d}{10 \, \mathrm{pc}} \right)^{-1}, \qquad (5.10)$$

thanks to the Earth's annual motion on its orbit. Finally, since the star–exoplanet system is moving relative to the solar system, it generally has a non-zero proper motion μ''. The parallactic motion, which has a distinctive period of one sidereal year, and the proper motion, which is constant on human timescales, can be readily identified and removed from the data.

Detecting an exoplanet by watching its parent star change its angular position on the sky is called the **astrometric method**. Equation 5.9 tells us that even for a massive exoplanet on a large orbit around a nearby star, measuring a_1'' requires sub-milliarcsecond imaging. This can be provided by astrometric satellites such as *Gaia*, or by radio interferometry, but it is difficult.

The systems for which a_1'' has been measured are all nearby. Thus, their distance $d \propto 1/p''$ is well known, and the stellar mass m_1 is usually fairly well known from the spectral and photometric properties of the parent star. The observed value of $\mathcal{P}$ tells us the star–exoplanet distance from Kepler's third law,

$$a = \left(\frac{\mathcal{P}}{2\pi} \right)^{2/3} \mu^{1/3} = \left(\frac{\mathcal{P}^2 G(m_1 + m_2)}{4\pi^2} \right)^{1/3}. \qquad (5.11)$$

We can then write

$$a_1 = \frac{m_2}{m_1 + m_2} a = m_2 \left(\frac{G\mathcal{P}^2}{4\pi^2(m_1 + m_2)^2} \right)^{1/3}. \qquad (5.12)$$

Knowing a_1, $\mathcal{P}$, and m_1, we can solve for m_2. In particular, when $m_2 \ll m_1$, this simplifies to

$$m_2 \approx a_1 \left(\frac{4\pi^2 m_1^2}{G\mathcal{P}^2} \right)^{1/3}. \qquad (5.13)$$

As an example, consider the cool dwarf star TVLM 513-46546, which has $p'' = 0.0932 \, \mathrm{arcsec}$, and thus is at a distance $d = 10.73 \, \mathrm{pc}$. Its mass is near the star / brown dwarf borderline, with $m_1 \approx 80 \, \mathrm{M_{jup}} \approx 0.08 \, \mathrm{M_\odot}$. Radio astrometry reveals that this star, in addition to having a proper motion $\mu'' = 0.0785 \, \mathrm{arcsec \, yr^{-1}}$, traces an ellipse with $a_1'' \approx 1.3 \times 10^{-4} \, \mathrm{arcsec}$. This corresponds to an orbital size $a_1 \approx 210\,000 \, \mathrm{km}$, about three times the radius of a typical cool dwarf star. The orbital period is $\mathcal{P} = 220 \, \mathrm{d}$. Using these values in Equation 5.13 yields $m_2 \approx 0.4 \, \mathrm{M_{jup}}$ for the mass of the unseen exoplanet. This in turn implies that the exoplanet's orbit has $a \approx 0.3 \, \mathrm{au}$.

An exoplanet always makes its parent star move on an elliptical orbit. However, the observed angular size a_1'' of the orbit will be, as we have seen, inconveniently small. Thus, it is generally easier to detect exoplanets by the radial velocity (RV) method, which relies on detecting the time-dependent Doppler shift of the parent star as it moves on its small orbit. Most stars have readily identifiable absorption lines in their spectra. If the central wavelength of an absorption line is λ_0 when the star is at rest relative to the observer, then it will be Doppler shifted by an amount $\Delta\lambda$ when the star has a radial velocity v_r relative to the observer. In the non-relativistic limit, the radial component of the star's velocity is given by the relation

$$v_r = c\frac{\Delta\lambda}{\lambda_0}. \tag{5.14}$$

Thus, when we repeatedly observe a star with an exoplanet, the star's orbital motion makes its value of v_r vary with a period $\mathcal{P}$ equal to the orbital period of the star–exoplanet system.[2]

Let's again take the simple case of a star on a circular orbit, with orbital speed given by Equation 5.8. As seen by an observer, the star's orbit is inclined by an angle i, as illustrated in Figure 5.1. If we define $t = 0$ to be at a moment when the radial velocity of the star has its largest value then, for a star on a circular orbit,

$$v_r(t) = v_{r,0} + v_1 \sin i \cos\left(\frac{2\pi t}{\mathcal{P}}\right), \tag{5.15}$$

where $v_{r,0}$ is the radial velocity of the barycenter of the star–exoplanet system. Thus, the radial velocity method can tell you $v_1 \sin i$ for a star on a circular orbit; if there is no astrometric information, the inclination i is usually unknown. Using Equation 5.8 for the circular speed v_1 of the star, and making use of Kepler's third law (Equation 5.11), we find that

$$v_1 = \frac{m_2}{m_1 + m_2}\left(\frac{G(m_1 + m_2)}{a}\right)^{1/3} = m_2\left(\frac{2\pi G}{\mathcal{P}(m_1 + m_2)^2}\right)^{1/3}. \tag{5.16}$$

In the limit $m_2 \ll m_1$, we can write

$$m_2 \sin i \approx v_1 \sin i \left(\frac{\mathcal{P}m_1^2}{2\pi G}\right)^{1/3}$$

$$\approx 0.352\,\mathrm{M_{jup}}\left(\frac{v_1 \sin i}{10\,\mathrm{m\,s^{-1}}}\right)\left(\frac{\mathcal{P}}{1\,\mathrm{yr}}\right)^{1/3}\left(\frac{m_1}{1\,\mathrm{M_\odot}}\right)^{2/3}. \tag{5.17}$$

Thus, if we know $\mathcal{P}$ and $v_1 \sin i$ from the radial velocity curve, and can make a good estimate of the stellar mass m_1, we can compute $m_2 \sin i$ for the exoplanet. This places a lower limit on the exoplanet's mass m_2.

---

[2] For ground-based observations, the observed v_r also varies with a period of one sidereal year and one sidereal day, thanks to the Earth's motion; these variations can be identified and removed because of their distinctive periods.

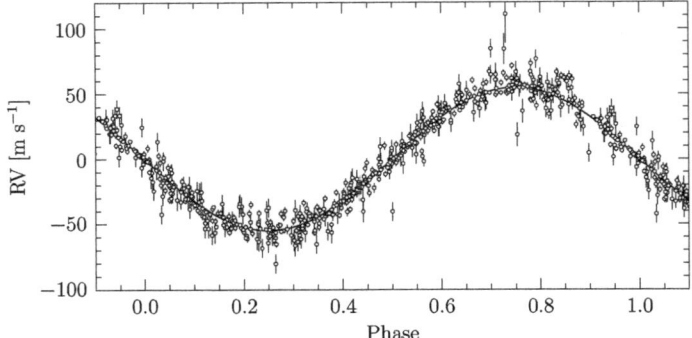

Figure 5.2 Radial velocity (RV) curve of 51 Pegasi. The data are folded over the period $\mathcal{P}$ = 4.230 786 9 d. The radial velocity $v_{r,0}$ = −33.2 km s^{-1} of the barycenter is removed from this plot. [Data compiled by Birkby *et al.* 2017]

The exoplanet 51 Pegasi b, also known as "Dimidium," was the first exoplanet discovered orbiting a main sequence star. It is also the prime example of an exoplanet on a nearly circular orbit found using the radial velocity method. The parent star, 51 Pegasi, is a Sun-like star with a mass m_1 = 1.05 M$_\odot$ and radius R_1 = 1.16 R$_\odot$. It is at a distance d = 15.53 pc and has a mean radial velocity $v_{r,0}$ = −33.2 km s^{-1}. However, in 1995 Michel Mayor and Didier Queloz announced that its radial velocity varied sinusoidally, as you would expect for a star on a circular orbit (see Figure 5.2). The orbital period of 51 Pegasi was found to be remarkably short: $\mathcal{P}$ = 4.23 d = 0.0116 yr, implying an exoplanet at a distance $a \approx 0.052$ au $\approx 10 R_1$. The amplitude of the sinusoidal variations was found to be quite large: $v_1 \sin i$ = 59 m s^{-1}. From Equation 5.17, this implies that 51 Pegasi b has $m_2 \sin i$ = 0.47 M$_{\text{jup}}$. Thus, 51 Pegasi b is the prototype of the class of **hot Jupiters**; these are exoplanets with masses comparable to or greater than that of Jupiter, on orbits smaller than that of Mercury ($a \sim 0.4$ au).

Although many exoplanets are on nearly circular orbits, others are on highly eccentric orbits, with $e > 0.5$. This means that the speed of the parent star varies along its orbit, with the maximum speed at periapsis exceeding the minimum speed at apoapsis by a factor

$$\frac{v_{1,\text{pe}}}{v_{1,\text{ap}}} = \frac{1 + e}{1 - e}. \tag{5.18}$$

When a star moves along an eccentric orbit, its observed radial velocity v_r does not vary as a perfect sinusoidal curve. Figure 5.3 demonstrates this by showing the radial velocity curves of two stars moving on eccentric orbits ($e \approx 0.75$) under the influence of an exoplanet. Figure 5.3(a) shows the radial velocity of the star HD 20868; this star's orbit is oriented so that the line of

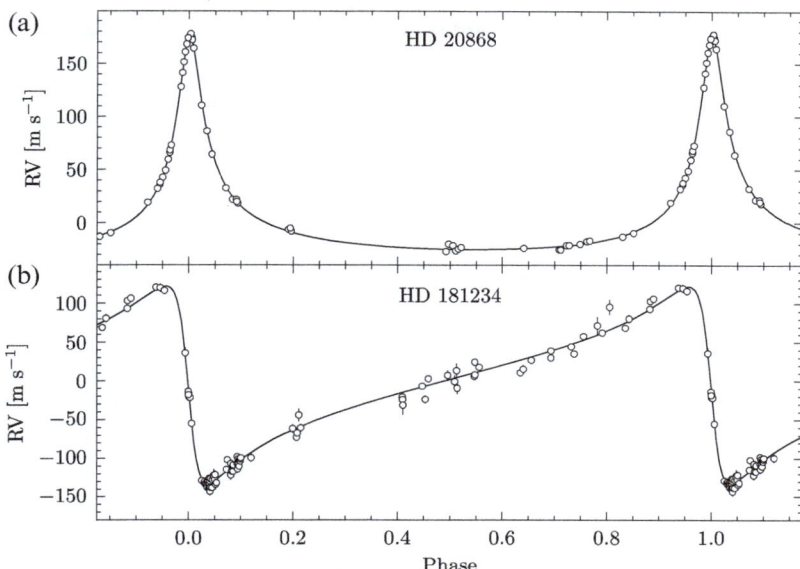

Figure 5.3 Phase-folded radial velocity curves of two eccentric systems; phase equals zero at periapsis. (a) Radial velocity curve of HD 20868: $\mathcal{P}$ = 380.8 d, e = 0.75, K = 100 m s^{-1}, ω_ℓ = 356°. [Moutou *et al.* 2009] (b) Radial velocity curve of HD 181234: $\mathcal{P}$ = 7462 d, e = 0.73, K = 127 m s^{-1}, ω_ℓ = 93°. [Rickman *et al.* 2019]

apsides is perpendicular to the observer's line of sight. Once per orbit, the star moves away from us with a relatively high speed when it is at periapsis ($v_{\rm pe} \sin i \approx 175$ m s^{-1}); half an orbit later, the star moves toward us with a relatively low speed when it is at apoapsis ($v_{\rm ap} \sin i \approx -25$ m s^{-1}). For comparison, Figure 5.3(b) shows the radial velocity of the star HD 181234; although this star's orbit has about the same eccentricity as that of HD 20868, its line of apsides is rotated by ∼90°. This means that at periapsis and apoapsis, the star's velocity is nearly perpendicular to the line of sight, and produces a very small Doppler shift. The observed radial velocity for HD 181234 changes rapidly near periapsis, when the star is strongly accelerated, and changes more slowly near apoapsis.

For a star on a Keplerian orbit with eccentricity e, the observed radial velocity can be written as

$$v_r(t) = v_{r,0} + K[\cos(v(t) + \omega_\ell) + e\cos\omega_\ell]. \qquad (5.19)$$

As before, $v_{r,0}$ is the radial velocity of the barycenter. The angle $v(t)$ is the true anomaly, as used in Equation 1.44; this measures how far the star has moved on its orbit relative to periapsis. In Equation 5.19, the angle ω_ℓ is the **longitude of periapsis**; this tells us the orientation of the line of apsides relative to the observer. The convention is that when the line of apsides is perpendicular to the observer's line of sight, $\omega_\ell = 0°$ when the star is moving away from the observer at periapsis, and $\omega_\ell = 180°$ when the star is moving toward the observer at periapsis. Finally, the velocity K is the **semi-amplitude** of the radial velocity; that is, half the difference between the maximum and minimum radial velocity. For a Keplerian orbit of arbitrary eccentricity, this is given by the relation

$$K = m_2 \sin i \left(\frac{2\pi G}{\mathcal{P}(m_1 + m_2)^2} \right)^{1/3} \frac{1}{(1 - e^2)^{1/2}}. \tag{5.20}$$

Suppose we have a long series of high-quality radial velocity measurements of a star; each measurement is tagged with the time when it was made. First, we correct for the Earth's rotation and orbital motion. Then we can use standard statistical methods for detecting periodicity to find the period $\mathcal{P}$ of the radial velocity variations.[3] If a single statistically significant period $\mathcal{P}$ is found, we then try to fit the radial velocity data with orbits of different eccentricity and orientation, varying e, ω_ℓ, K, and $v_{r,0}$ until we find the best fit.

Once we know $\mathcal{P}$, K, and e for the best-fitting orbit, we can use Equation 5.20 in the limit $m_2 \ll m_1$ to find

$$m_2 \sin i \approx K(1 - e^2)^{1/2} \left(\frac{\mathcal{P} m_1^2}{2\pi G} \right)^{1/3} \tag{5.21}$$

$$\approx 0.35 \, \mathrm{M_{jup}}(1 - e^2)^{1/2} \left(\frac{K}{10 \, \mathrm{m \, s^{-1}}} \right) \left(\frac{\mathcal{P}}{1 \, \mathrm{yr}} \right)^{1/3} \left(\frac{m_1}{1 \, \mathrm{M_\odot}} \right)^{2/3}.$$

To find $m_2 \sin i$ from the radial velocity data, we thus need to know the stellar mass m_1 by some independent method, in addition to finding $\mathcal{P}$, K, and e from the parent star's radial velocity variations.

5.2 Transit Detection

Another method of detecting exoplanets is based on the facts that (i) exoplanets are opaque and (ii) stars emit light. The **transit method** involves monitoring the flux of light from a star and looking for periodic intervals of dimming as the exoplanet passes in front of the star, from the observer's viewpoint. To help in understanding the geometry of transits, Figure 5.4 shows a system consisting of an exoplanet on a circular orbit about its parent star. For visualization purposes, the exoplanet shown has a large radius ($R_2 = 0.2R_1$) and is on a small orbit ($a = 6R_1$); think of it as a puffy ultra-hot Jupiter. An observer at a distance d sees the exoplanet orbit inclined by an angle $0 \le i \le 90°$ relative to

---

[3] A Lomb–Scargle periodogram, for instance, is often used for finding periodic signals in data sampled at irregular intervals.

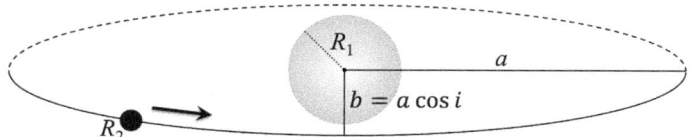

Figure 5.4 A star with radius R_1 is orbited by an exoplanet with radius $R_2 = 0.2R_1$; the radius of the circular orbit is $a = 6R_1$. The orbit is tilted by an angle $i \approx 78.5°$ relative to the plane of the sky. The half of the orbit farther from the observer is shown as a dashed line.

the plane of the sky. The circular orbit is thus seen in projection as an ellipse with semimajor axis $a'' = a/d$ and semiminor axis $b'' = a \cos i/d$.

The star has an angular radius $\theta_1 = R_1/d$; similarly, the exoplanet has an angular radius $\theta_2 = R_2/d$. As seen by the observer, the exoplanet's center is closest to the star's center when it is at the point on its orbit closest to the observer. The projected separation between the exoplanet's center and the star's center at this point, called the impact parameter, is equal to the semiminor axis of the projected ellipse, $b'' = a \cos i/d$. The requirement for a **grazing transit**, with the exoplanet's disk just barely touching the star's disk, is then

$$b'' = \theta_1 + \theta_2. \tag{5.22}$$

This can be simplified to a requirement for the inclination,

$$\cos i = \frac{R_1 + R_2}{a} \qquad \text{[grazing]}. \tag{5.23}$$

Figure 5.4 shows a system that has a grazing transit once per orbit.

The requirement for a **full transit**, with the exoplanet's disk falling entirely within the star's disk, is

$$b'' < \theta_1 - \theta_2 \tag{5.24}$$

or

$$\cos i < \frac{R_1 - R_2}{a} \qquad \text{[full]}. \tag{5.25}$$

Scaling to the Earth transiting the Sun, with $R_2 = 0.009\,16\,R_\odot$ and $a = 1$ au $= 215\,R_\odot$, the inclination requirement for a full transit becomes

$$\cos i < 0.004\,61 \left(\frac{R_1 - R_2}{0.9908\,R_\odot}\right)\left(\frac{a}{1\,\text{au}}\right)^{-1}. \tag{5.26}$$

Thus, a distant observer could see the Earth fully transit the Sun only if the inclination were $i > 89.74°$. To put this another way, only observers within $0.26°$ of the ecliptic would be able to detect the Earth by the transit method; the fraction of the celestial sphere covered by this narrow strip is $f = 0.004\,61$.

(The exoplanet K2-65 b, at a distance $d = 63$ pc, is among the closest "transit buddies" of the Earth. As seen from the Earth, it transits its parent star K2-65; in addition, since the K2-65 system has an ecliptic latitude $\beta = 0.12°$, the Earth transits the Sun as seen from K2-65 b.)

If we scale to Jupiter transiting the Sun, with $R_2 = 0.100\,R_\odot$ and $a = 5.2\,\mathrm{au} = 1120\,R_\odot$, we find the requirement for a full transit is

$$\cos i < 8.0 \times 10^{-4} \left(\frac{R_1 - R_2}{0.90\,R_\odot} \right) \left(\frac{a}{5.2\,\mathrm{au}} \right)^{-1}. \tag{5.27}$$

Thus, for a distant observer to see Jupiter fully transit the Sun, the inclination would have to be $i > 89.95°$. If the exoplanet Dimidium, on its $a = 0.052$ au orbit, is comparable in size to Jupiter, we would see it fully transit its parent star 51 Pegasi if the orbital inclination were $\cos i < 0.095$, or equivalently $i > 84.6°$. Alas, we do not see Dimidium transit 51 Pegasi. However, since the upper limit on $\cos i$ for transits is inversely proportional to a, exoplanets on smaller orbits are statistically more likely to be seen transiting their star.

From the observer's viewpoint, the projected angular areas of the exoplanet and star are $\pi\theta_2^2$ and $\pi\theta_1^2$. Thus, if the exoplanet fully transits the star, the fraction of the star's projected area that it covers is

$$\frac{dA}{A} = \left(\frac{\theta_2}{\theta_1} \right)^2 = \left(\frac{R_2}{R_1} \right)^2. \tag{5.28}$$

Scaling to the Earth or Jupiter transiting the Sun, the covering fraction is

$$\frac{dA}{A} = 8.39 \times 10^{-5} \left(\frac{R_2}{1\,R_\oplus} \right)^2 \left(\frac{R_1}{1\,R_\odot} \right)^{-2}$$
$$= 1.01 \times 10^{-2} \left(\frac{R_2}{1\,R_{\mathrm{jup}}} \right)^2 \left(\frac{R_1}{1\,R_\odot} \right)^{-2}. \tag{5.29}$$

With current technology, the disks of exoplanets cannot be readily resolved in angle; even as seen from our close neighbor, Proxima Centauri, the angular diameter of Jupiter is $2\theta_2 \sim 7 \times 10^{-4}$ arcsec. However, it is possible to monitor the flux F of a planet to see whether there are periodic dips in flux due to the transit of an exoplanet. If the surface brightness of a star were uniform over its disk, then the fractional dip in flux, dF/F, would be equal to the fraction of the star's area covered, dA/A. Equation 5.29 then tells us to expect $dF/F \sim 0.01$ for gas giant exoplanets transiting Sun-like stars; measuring 1% flux variations is fairly straightforward. However, we expect $dF/F \sim 10^{-4}$ for terrestrial exoplanets transiting Sun-like stars; measuring 0.01% flux variations is challenging; with current technology, it requires dedicated satellites.

To illustrate some of the physical properties that can be measured for a star–exoplanet system, Figure 5.5(a) shows a transiting system. For this system, we

(a)

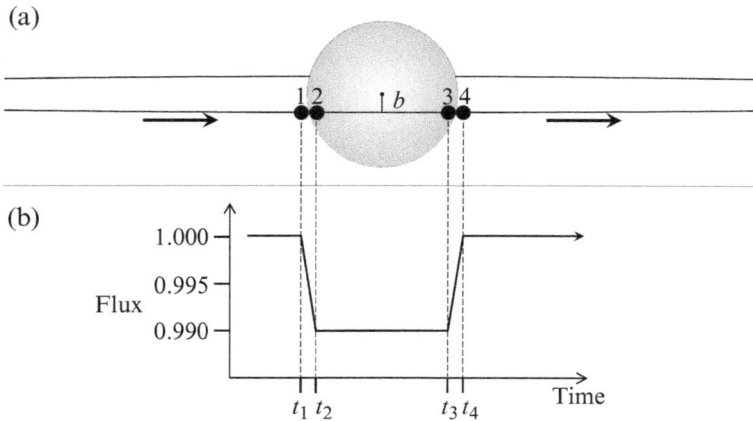

(b)

Figure 5.5 (a) An exoplanet with $R_2 = 0.1R_1$ transits a star on an orbit with $a = 10R_1$ and $\cos i = 0.025$. (b) The resulting light curve, normalized to a stellar flux $F = 1$ in the absence of a transit. Uniform surface brightness is assumed for the star.

have chosen $R_2 = 0.1R_1$ and $a = 10R_1$ to mimic a hot Jupiter orbiting a Sun-like star. The assumed inclination is $\cos i = 0.025$, or equivalently $i = 88.6°$. The impact parameter b, expressed in physical rather than angular units, is $b = a \cos i$; for the illustrated system, $b = 0.25R_1 = 2.5R_2$.

Figure 5.5(b) shows a schematic light curve for the star during the exoplanet's transit, showing the decrease in flux $dF/F = (R_2/R_1)^2 = 0.01$ expected in mid-transit. The light curve is schematic partly because it ignores the intrinsic variability in the star's luminosity; even a sedate star like the Sun has intrinsic variability on a level $dF/F \sim 0.001$. Although the dips resulting from transits have a distinct shape compared to intrinsic stellar variability, it is prudent to wait for multiple successive transits at a constant interval $\mathcal{P}$ before crying, "Eureka! I have found an exoplanet!" From the transit depth dF/F, we compute the ratio of the exoplanet's radius to the star's radius,

$$\frac{R_2}{R_1} = \left(\frac{dF}{F}\right)^{1/2}. \tag{5.30}$$

In addition, the interval from one transit to the next is equal to $\mathcal{P}$, the orbital period of the system.

If the exoplanet is on a circular orbit, as we assume in Figure 5.5, its speed relative to its parent star is

$$v = \frac{2\pi a}{\mathcal{P}} = \left(\frac{2\pi\mu}{\mathcal{P}}\right)^{1/3}, \tag{5.31}$$

making use of Kepler's third law. As with the radial velocity method, the star's spectrum usually lets us deduce m_1. As long as the star isn't near the star /

brown dwarf border, it's reasonable to assume that $\mu \equiv G(m_1 + m_2) \approx Gm_1$. This means we can deduce the exoplanet's orbital speed from its period $\mathcal{P}$:

$$v \approx 98.9 \, \mathrm{km \, s}^{-1} \left(\frac{m_1}{1 \, M_\odot} \right)^{1/3} \left(\frac{\mathcal{P}}{10 \, \mathrm{d}} \right)^{-1/3}. \tag{5.32}$$

Knowing the velocity is useful since it lets us translate the duration of different transit stages into physical distances. For instance, in the limit of a perfectly edge-on orbit, we expect the duration of the transit to be comparable to the time it takes the exoplanet to move a distance equal to the star's diameter,

$$\tau_0 \equiv \frac{2R_1}{v} \approx 3.911 \, \mathrm{hr} \left(\frac{R_1}{1 \, \mathrm{R}_\odot} \right) \left(\frac{m_1}{1 \, M_\odot} \right)^{-1/3} \left(\frac{\mathcal{P}}{10 \, \mathrm{d}} \right)^{1/3}. \tag{5.33}$$

We therefore expect a transit to last for a small fraction of an orbital period. For instance, if a hot Jupiter with $\mathcal{P} \approx 20 \, \mathrm{d}$ orbits a Sun-like star, it will have $\tau_0 \approx 0.01\mathcal{P}$. The brevity of transits makes them easy to miss (especially if you are observing from a cloudy planet). However, it also means that you can simplify the geometry of a transit by assuming the exoplanet moves in a straight line at a constant speed while transiting.

Consider, once again, the transit illustrated in Figure 5.5. From the observed light curve, we can identify four critical times during the transit. First contact (time t_1) occurs when the exoplanet first touches the star from the observer's viewpoint. Second contact (t_2) occurs when the exoplanet's disk first lies entirely within the star's disk. Near the end of the transit, third contact (t_3) is the last instant when the exoplanet's disk lies entirely within the star's disk. The transit ends with fourth contact (t_4), when the exoplanet last touches the star from the observer's viewpoint. If the exoplanet passes exactly through the center of the star's disk ($b = 0$), then the time between first and third contact, or between second and fourth contact, is exactly equal to the scaling time τ_0:

$$t_3 - t_1 = t_4 - t_2 = \tau_0 = \frac{2R_1}{v}. \tag{5.34}$$

Thus, knowing the times of contact and the orbital period, we can compute the radius of the star as

$$R_1 = \frac{v(t_3 - t_1)}{2} \approx 1.023 \, \mathrm{R}_\odot \left(\frac{m_1}{1 \, M_\odot} \right)^{1/3} \left(\frac{\mathcal{P}}{10 \, \mathrm{d}} \right)^{-1/3} \left(\frac{t_3 - t_1}{4 \, \mathrm{hr}} \right), \tag{5.35}$$

making use of Equation 5.32 for the speed v of the exoplanet. Since our computed value of R_1 is proportional to $m_1^{1/3}$, we can find the mean density of the star,

$$\bar{\rho}_1 \equiv \frac{3m_1}{4\pi R_1^3} \approx 1.32 \, \mathrm{g \, cm}^{-3} \left(\frac{\mathcal{P}}{10 \, \mathrm{d}} \right) \left(\frac{t_3 - t_1}{4 \, \mathrm{hr}} \right)^{-3}, \tag{5.36}$$

purely from the observed transit properties, without having to know m_1 from independent methods.

We can also use transit measurements to find the radius of the exoplanet. The **ingress time**, $t_2 - t_1$, is equal to the **egress time**, $t_4 - t_3$, in the case of a circular orbit. When $b = 0$, the ingress time is the time it takes the exoplanet to travel a distance equal to its own diameter:

$$t_2 - t_1 = \frac{2R_2}{v} = \tau_0 \left(\frac{R_2}{R_1} \right). \tag{5.37}$$

Thus, knowing the ingress time and orbital period, we can compute the radius of the exoplanet as

$$R_2 = \frac{v(t_2 - t_1)}{2} \approx 0.848 \, \mathrm{R_{jup}} \left(\frac{m_1}{1 \, \mathrm{M_\odot}} \right)^{1/3} \left(\frac{P}{10 \, \mathrm{d}} \right)^{-1/3} \left(\frac{t_2 - t_1}{20 \, \mathrm{min}} \right). \tag{5.38}$$

From the radius of the star (Equation 5.35) and the exoplanet (Equation 5.38), we can compute a predicted dip in flux during full transit,

$$\left(\frac{dF}{F} \right)_{\mathrm{pred}} = \left(\frac{t_2 - t_1}{t_3 - t_1} \right)^2 = 6.9 \times 10^{-3} \left(\frac{t_2 - t_1}{20 \, \mathrm{min}} \right)^2 \left(\frac{t_3 - t_1}{4 \, \mathrm{hr}} \right)^{-2}. \tag{5.39}$$

If this prediction doesn't match the observed dip in the star's flux, then we must go back and scrutinize our assumptions. In particular, we should be sceptical of the assumption that the transit has $b = 0$; that is, that the exoplanet passes *exactly* through the center of the star's disk.

It is not always true that the impact parameter b is negligibly small compared to the stellar radius R_1. We can, however, determine the impact parameter from the observed properties of the transit. We start with the approximation that the transit path is a straight line and the speed of the exoplanet has a constant value v. We then recognize that the impact parameter b, representing the shortest distance from the transit path to the center of the star's disk, must be perpendicular to the transit path (Figure 5.5). The Pythagorean theorem then tells us that the complete duration of the transit, from first to fourth contact, is

$$t_4 - t_1 = \frac{2}{v} \left[(R_1 + R_2)^2 - b^2 \right]^{1/2} = \tau_0 \left[(1 + R_2/R_1)^2 - b^2/R_1^2 \right]^{1/2}. \tag{5.40}$$

Similarly, the time that elapses between the end of ingress and the start of egress (that is, the time when the transit is full) is

$$t_3 - t_2 = \tau_0 \left[(1 - R_2/R_1)^2 - b^2/R_1^2 \right]^{1/2}. \tag{5.41}$$

We can define a dimensionless number

$$f_{\mathrm{full}} \equiv \frac{t_3 - t_2}{t_4 - t_1}, \tag{5.42}$$

which represents the fraction of the complete transit (from first to fourth contact) that is a full transit, with the exoplanet's disk entirely within the star's disk. Using Equations 5.40 and 5.41, we find

$$f_{\text{full}}^2 = \frac{(1 - R_2/R_1)^2 - b^2/R_1^2}{(1 + R_2/R_1)^2 - b^2/R_1^2}. \tag{5.43}$$

Since we know f_{full} and R_2/R_1 from the duration and shape of the transit light curve, we can solve for b (in units of the stellar radius R_1):

$$\frac{b^2}{R_1^2} = \frac{(1 - R_2/R_1)^2 - f_{\text{full}}^2(1 + R_2/R_1)^2}{1 - f_{\text{full}}^2}. \tag{5.44}$$

Suppose you observe a transit with $dF/F = 0.01$ and thus $R_2/R_1 = 0.1$. If the orbiting exoplanet has a perfectly edge-on circular orbit, then you will measure $f_{\text{full}} = 0.9/1.1 = 0.818$, with ingress and egress each accounting for 9.1% of the complete transit time. However, if we observed the transit illustrated in Figure 5.5, we would measure $dF/F = 0.01$ and $f_{\text{full}} = 0.807$, then deduce that $R_2/R_1 = 0.1$ and $b/R_1 = 0.25$ for this particular system. Knowing the impact parameter lets us make the required geometric corrections for properties such as the exoplanet's radius. For modest values of b/R_1, the necessary corrections are (compared to Equations 5.35 and 5.38):

$$R_1 \approx \frac{v(t_3 - t_1)}{2} \left[1 + \frac{1}{2} \frac{b^2}{R_1^2} \right], \tag{5.45}$$

$$R_2 \approx \frac{v(t_2 - t_1)}{2} \left[1 - \frac{1}{2} \frac{b^2}{R_1^2} \right]. \tag{5.46}$$

Ignoring this correction would lead us to underestimate the star's radius, overestimate the star's density, and overestimate the exoplanet's radius.

Because of the geometry required, the vast majority of exoplanets discovered by the transit method have $\cos i < 0.2$. This implies that the typical transiting exoplanet has $\sin i > 0.98$. If we shift our attention from the transiting exoplanet to its parent star, we note that the semi-amplitude of the star's radial velocity curve is $K \propto P^{-1/3} m_2 \sin i$, from Equation 5.20. Thus, the same factors that make a transit more likely to be detected – a gas giant exoplanet on an edge-on orbit with a short period – also increase the value of K, making the Doppler shifts easier to measure.

Consider an exoplanet that is discovered by the transit method and is then subjected to follow-up by the radial velocity method. Its radius R_2 is known from the transits (Equation 5.46) and its mass m_2 is known from the radial velocity variations of its parent star (Equation 5.22). Figure 5.6 shows the radius and mass for a sample of exoplanets with high-quality data. The curved lines in Figure 5.6 show the expected mass–radius relation for cool spheres of different composition (pure iron, rock, water, or hydrogen). At

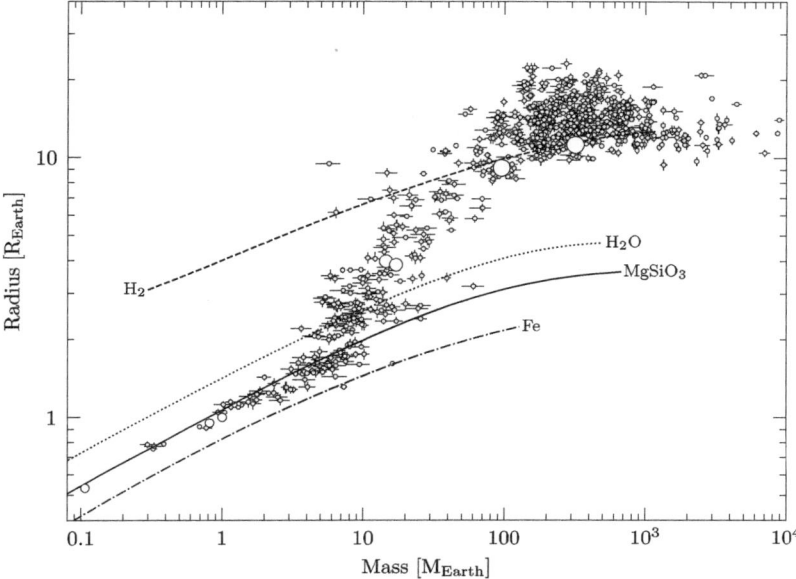

Figure 5.6 Radius versus mass for exoplanets with transit and radial velocity data. Solar system planets are shown as large open circles (with apologies to Mercury). The radius–mass relation for cool self-gravitating spheres is shown for iron (Fe), the mineral enstatite ($MgSiO_3$), water (H_2O), and hydrogen (H_2).

low masses, compression by self-gravity is small, and the radius grows as $R \propto m^{1/3}$. At higher masses, compression becomes important; in fact, at a mass $m \sim 1200\,M_\oplus$, with some dependence on chemical composition, the radius reaches a maximum. At higher masses, the exoplanet is supported primarily by electron degeneracy pressure, with $R \propto m^{-1/3}$. The tools provided by Newtonian celestial dynamics let us find the mass and radius of an exoplanet; although this doesn't permit us to determine the interior structure of a multilayered exoplanet, it does give useful constraints.

5.3 Transit Timing Variations

A perfect Keplerian system, consisting of two spherical objects orbiting their mutual center of mass, will have a constant orbital period $\mathcal{P}$. If the secondary of such a system transits the primary from our point of view, then the transits will occur separated by a constant interval $\mathcal{P}$. However, no two-body system is a perfect Keplerian system. The period $\mathcal{P}$ can change because of tidal effects between the two slightly non-spherical objects. It can also change because of the disturbing force of additional objects in the system. Since the intervals between transits can be measured with high precision, the detection of **transit timing variations** (TTVs) is a useful tool for determining the tidal evolution of systems containing a transiting exoplanet, and for detecting additional (non-transiting) exoplanets in the system.

Suppose that a transiting exoplanet has been monitored long enough that several of its transits have been detected. For each individual transit, the time of mid-transit is measured. One particular transit is designated as "transit zero"; the midpoint of transit zero occurs at a time t_0. If the orbital period $\mathcal{P}$ of the exoplanet is constant, then the time of mid-transit for earlier or later transits can be calculated as

$$C = t_0 + \mathcal{P} \cdot E, \tag{5.47}$$

where the transit number[4] E is the integer assigned to each transit: $E = 0$ for transit zero, $E = 1$ for the first transit after transit zero, and so forth. However, if our assumption of constant $\mathcal{P}$ is incorrect, there will be significant residuals in $O - C$, where $C(E)$ is the calculated time of transit from Equation 5.47 and O is the actually observed time of transit.

To see how TTVs can be a diagnostic for the tidal evolution of a two-body system, let's take an exaggerated example. Suppose we observe five successive transits, at epochs we define as $E = -2, -1, 0, 1$, and 2. We measure the time between successive transits as 1.03 d, 1.01 d, 0.99 d, and 0.97 d. The mean period that we find is therefore $\mathcal{P} = 1.00$ d. However, the calculated times of transit,

$$C - t_0 = -2.00\,\text{d}, \quad -1.00\,\text{d}, \quad 0.00\,\text{d}, \quad +1.00\,\text{d}, \quad +2.00\,\text{d}, \tag{5.48}$$

differ from the observed times of transit,

$$O - t_0 = -2.04\,\text{d}, \quad -1.01\,\text{d}, \quad 0.00\,\text{d}, \quad +0.99\,\text{d}, \quad +1.96\,\text{d}, \tag{5.49}$$

by an amount

$$O - C = -0.04\,\text{d}, \quad -0.01\,\text{d}, \quad 0.00\,\text{d}, \quad -0.01\,\text{d}, \quad -0.04\,\text{d} \tag{5.50}$$

that is quadratic in the transit number E.

Although this example has an implausibly swift decrease in $\mathcal{P}$, a much smaller – but still measurable – rate of decrease can result from tidal evolution in a system containing an ultra-hot Jupiter with $\mathcal{P} < 2$ d. Mature Sun-like stars generally have rotation periods $\mathcal{P}_{\text{rot}} > 2$ d, so we expect ultra-hot Jupiters to spiral inward, like Phobos around Mars, with decreasing a and $\mathcal{P}$. One example of a transiting exoplanet on a decaying orbit is the ultra-hot Jupiter WASP-12 b. From the light curves of its transits, it is deduced to be on a small, nearly circular orbit with $a = 3.06 R_1$ and $\cos i \approx 0.12$. Using data taken from 2019 Dec to 2021 Dec, the best-fitting orbital period was $\mathcal{P} = 1.091\,417$ d. However, looking over a longer interval from 2009 to 2021, a plot of $O - C$

---

[4] The transit number is also known as the "epoch"; hence the use of the letter E to symbolize it.

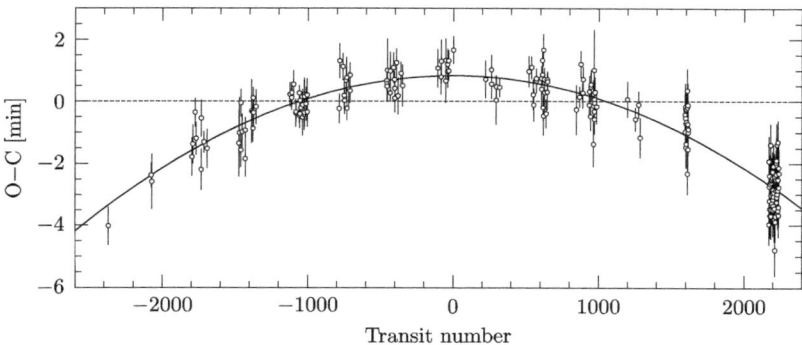

Figure 5.7 Observed minus calculated transit times for WASP-12 b, assuming a constant period $\mathcal{P}$. [Wong *et al.* 2022]

(assuming constant $\mathcal{P}$) shows the parabolic curve of Figure 5.7. The downward "frown" of the $O - C$ curve indicates that the true period is decreasing with time; an upward "smile" would be the indicator of a period that increases with time.

For a system with significant tidal evolution, like the WASP-12 system, we can usefully make a quadratic calculation for the time of mid-transit,

$$C_{\text{quad}} = t_0 + \mathcal{P} \cdot E + \frac{1}{2}\frac{d\mathcal{P}}{dE} \cdot E^2, \tag{5.51}$$

where $\mathcal{P}$ is the orbital period at the time of transit zero and $d\mathcal{P}/dE$, assumed to be constant in this approximation, represents the change in orbital period between one transit and the next. For WASP-12 b, the data displayed in Figure 5.7 are best fitted with $\mathcal{P} = 1.091\,419\,4$ d and $d\mathcal{P}/dE = -1.03 \times 10^{-9}$ d. Although this value of $d\mathcal{P}/dE$ represents a decrease of only 89 μs per orbit, in Equation 5.51 it enters the term proportional to E^2; thus, at an epoch $E \approx 2000$, the timing deviation from the constant-period result is $\sim$3 min.

Let's see what the observed value of $d\mathcal{P}/dE$ can tell us about the tidal properties of a system. From Equation 4.56, the tidal evolution timescale is

$$t_{\text{TE}} = \frac{1}{6\pi}\frac{Q}{k_{\text{T}}}\frac{m_1}{m_2}\left(\frac{a}{R_1}\right)^5 \mathcal{P}. \tag{5.52}$$

Using Equation 4.59, we can relate the evolution timescale to the change in period per orbit, $d\mathcal{P}/dE$:

$$t_{\text{TE}} = \frac{3}{2}\mathcal{P}\left|\frac{d\mathcal{P}}{dt}\right|^{-1} = \frac{3}{2}\mathcal{P}^2\left|\frac{d\mathcal{P}}{dE}\right|^{-1}. \tag{5.53}$$

Combining Equations 5.52 and 5.53, we can find the value of Q/k_{T} for a tidally disturbed parent star from properties determined by tidal timing variations and radial velocity measurements:

$$\frac{Q}{k_T} = 9\pi \mathcal{P} \left| \frac{d\mathcal{P}}{dE} \right|^{-1} \frac{m_2}{m_1} \left(\frac{a}{R_1} \right)^{-5} \tag{5.54}$$

$$= 1.16 \times 10^5 \left(\frac{\mathcal{P}}{1\,\mathrm{d}} \right) \left(\frac{|d\mathcal{P}/dE|}{10^{-9}\,\mathrm{d}} \right)^{-1} \left(\frac{m_2/m_1}{0.001} \right) \left(\frac{a/R_1}{3} \right)^{-5}.$$

We have scaled our results to a system like WASP-12, which we know from radial velocity measurements has $m_2/m_1 = 9.8 \times 10^{-4}$, very similar to the Jupiter/Sun mass ratio.

For WASP-12 b, the value of $Q/k_T \approx 1.1 \times 10^5$ that we deduce is much larger than the values $Q/k_T \sim 40$ found for the Earth–Moon system and $Q/k_T \sim 500$ found for the Mars–Phobos system. As mentioned in Section 4.4, this results from a combination of a low tidal Love number k_T for the centrally concentrated star and a large tidal quality factor Q for the relatively inviscid gas of which the star is made. Despite the large value of Q/k_T for WASP-12 b, its calculated tidal evolution time is, from Equation 5.53, only $t_{TE} = 4.75\,\mathrm{Myr}$. Since the estimated age of the WASP-12 system is $t \sim 2\,\mathrm{Gyr}$, this means we are catching WASP-12 b in the last $\sim 0.2\%$ of its existence before it spirals into its parent star.

Exoplanets on larger orbits around their parent star, in contrast to ultra-hot Jupiters such as WASP-12 b, do not show a steady decrease in their orbital period $\mathcal{P}$. However, they can show fluctuations on short timescales due to the perturbative gravitational effect of other exoplanets in their system. If the other exoplanets do not transit, then the TTVs they produce in their transiting neighbor provide an indirect means of detecting them. As an example, Figure 5.8 shows the values of $O - C$, assuming a constant orbital period $\mathcal{P}$, for two exoplanets in different systems. The exoplanet Kepler-419 b, Figure 5.8(a), has a mean orbital period $\mathcal{P} = 69.755\,\mathrm{d}$. However, individual transits have a variation $\delta t = \langle (O - C)^2 \rangle^{1/2} \sim 0.9\,\mathrm{hr} \sim 0.0005\mathcal{P}$ about the predicted time C. Its plot of $O - C$ versus transit number E doesn't show strongly periodic behavior during the time it was surveyed. For comparison, the exoplanet Kepler-88 b, Figure 5.8(b), has a mean orbital period $\mathcal{P} = 10.916\,\mathrm{d}$, with individual transits arriving with variation $\delta t \sim 8\,\mathrm{hr} \sim 0.03\mathcal{P}$. Here, the curve of $O - C$ versus E is more strongly periodic, on a timescale $\sim 55\mathcal{P} \sim 600\,\mathrm{d}$.

A transiting exoplanet usually inhabits a planetary system with other orbiting exoplanets. The disturbing function $\mathcal{R}$ provided by each of the neighboring exoplanets varies in a complicated way, dependent on the orbital parameters of both the transiting exoplanet and the perturbing exoplanet. To get a feel for the expected amplitude of TTVs, let's look at a simple three-body system. The primary is a star of mass m_1; it is transited by an exoplanet on a nearly circular orbit with semimajor axis a_{tran} and orbital period $\mathcal{P}_{tran}$. The perturbing third body is an exoplanet of mass $m_{pert} < m_1$ whose orbit is nearly

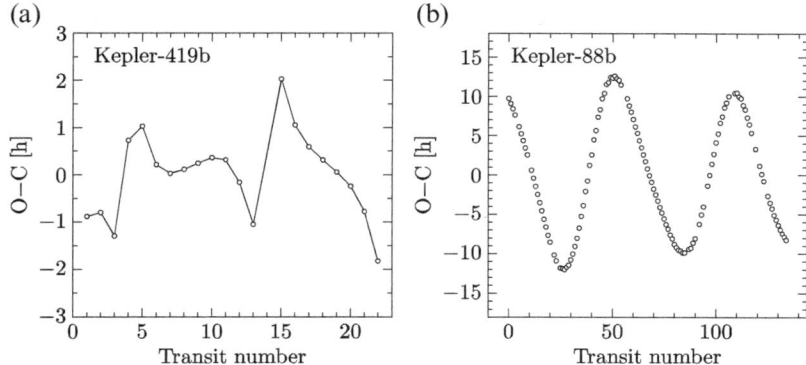

Figure 5.8 (a) Transit timing variations for the exoplanet Kepler-419 b ($\mathcal{P}$ = 69.755 d). [Dawson *et al.* 2014] (b) Transit timing variations for the exoplanet Kepler-88 b ($\mathcal{P}$ = 10.916 d). [Holczer *et al.* 2016]

coplanar with that of the transiting exoplanet, but which has $a_{\text{pert}} > a_{\text{tran}}$ and thus $\mathcal{P}_{\text{pert}} > \mathcal{P}_{\text{tran}}$. What can we say about the variance in transit times, $\delta t = \langle (O - C)^2 \rangle^{1/2}$, created by the outer perturbing object?

The transit times vary because the perturbing object accelerates the transiting exoplanet. Since the gravitational acceleration of an object is independent of its mass, we expect δt to depend on the perturber's mass m_{pert}, but not on the transiting exoplanet's mass. Moreover, the transit times will be affected only if the transiting exoplanet is accelerated relative to its parent star. Since a differential acceleration is required, we can think of TTVs, in this case, as being a tidal effect that falls as the cube of the distance. Thus, we can write a dimensionless TTV in a similar form to the dimensionless tidal amplitude ε_{T} (Equation 4.15):

$$\frac{\delta t}{\mathcal{P}_{\text{tran}}} \sim \frac{m_{\text{pert}}}{m_1} \left(\frac{a_{\text{tran}}}{a_{\text{pert}}} \right)^3 . \tag{5.55}$$

Using Kepler's third law, we can write this in the form

$$\frac{\delta t}{\mathcal{P}_{\text{tran}}} \sim \frac{m_{\text{pert}}}{m_1} \left(\frac{\mathcal{P}_{\text{tran}}}{\mathcal{P}_{\text{pert}}} \right)^2 . \tag{5.56}$$

This is purely a back-of-the-envelope result, but gives an idea of what to expect in the limit of a perturber on a nearly circular orbit with $\mathcal{P}^2_{\text{pert}} \gg \mathcal{P}^2_{\text{tran}}$. For instance, a hypothetical observer on K2-65 b, observing transits of the Earth, would detect TTVs of order $\delta t \sim 7 \times 10^{-6}$ yr ~ 4 min from Jupiter's perturbative influence.

One way to determine the properties of a perturbing exoplanet is to do numerical three-body simulations, varying the mass and orbital properties of the perturbing exoplanet until a good match is found to the observed transit times. For instance, the observed transits of Kepler-419 b, as shown in Figure 5.8(a), are consistent with a perturber that has $m_{pert} \approx 7.6\,M_{jup}$, on an orbit with $\mathcal{P}_{pert} \approx 670\,d \approx 9.7\mathcal{P}_{tran}$. Given that the mass of the parent star in the system is $m_1 \approx 1.4\,M_\odot \approx 190m_{pert}$, Equation 5.56 leads us to expect TTVs of order $\delta t \sim 6 \times 10^{-5}\mathcal{P}_{tran} \sim 0.1\,d$. This is an order of magnitude smaller than the observed variations. In the case of the Kepler-419 system, the transit timing variations are enhanced by the fact that the transiting exoplanet is on a highly eccentric orbit, with $e = 0.83$. This leads to an increased perturbation whenever the transiting exoplanet is dawdling near apoapsis.

An additional way to increase δt is to have a perturber in **resonance** with the transiting exoplanet. Let us now suppose that the outer perturbing object is nearly in a mean motion resonance with the transiting exoplanet, with

$$\frac{\mathcal{P}_{pert}}{\mathcal{P}_{tran}} \approx \frac{\alpha}{\beta}, \tag{5.57}$$

where α and β are mutually prime positive integers. The deviation of the system from a perfect resonance is given by the dimensionless parameter

$$\Delta \equiv \frac{\mathcal{P}_{pert}}{\mathcal{P}_{tran}} \frac{\beta}{\alpha} - 1. \tag{5.58}$$

As an example, the transiting exoplanet Kepler-88 b has $\mathcal{P}_{tran} = 10.916\,d$. Its transit timing variations, shown in Figure 5.8(b), can be explained by the presence of a perturbing planet (Kepler-88 c) with $\mathcal{P}_{pert} = 22.267\,d$. These two exoplanets are thus close to a 2 : 1 resonance, with $\mathcal{P}_{pert}/\mathcal{P}_{tran} = 2.040$ and $\Delta \approx 0.020$.

Notice that if a transiting planet and its outer perturber are relatively closely spaced, with $1 < \mathcal{P}_{pert}/\mathcal{P}_{tran} < 2.3$, they can never be far from a **first-order resonance**. (A first-order resonance is one with $\beta = \alpha - 1$. Thus, the first-order resonances are 2 : 1, 3 : 2, 4 : 3, and so on.) For such a closely spaced pair of planets, the value of Δ for the nearest first-order resonance will always have $|\Delta| < 0.15$. If the pair of orbits is close to a resonance, the value of δt, the typical TTV of the inner planet, is given by

$$\frac{\delta t}{\mathcal{P}_{tran}} \sim \frac{m_{pert}}{m_1} \frac{1}{|\Delta|}. \tag{5.59}$$

For example, the Kepler-88 system has $1/\Delta \approx 50$; this gives more than a hundred-fold enhancement of $\delta t/\mathcal{P}_{tran}$ relative to the value you would calculate from Equation 5.56 in the absence of a resonance.

The periodicity of the TTVs in the Kepler-88 system is also related to the fact that there is a near-resonance in the system. The synodic period of an inner transiting planet and an outer perturbing planet is given by

$$\frac{1}{\mathcal{P}_{\text{syn}}} = \frac{1}{\mathcal{P}_{\text{tran}}} - \frac{1}{\mathcal{P}_{\text{pert}}}. \tag{5.60}$$

For planets near the $\alpha : \beta$ resonance, this can be written as

$$\mathcal{P}_{\text{syn}} = \mathcal{P}_{\text{tran}}\frac{\alpha(1 + \Delta)}{\alpha(1 + \Delta) - \beta}. \tag{5.61}$$

In the more specific case of the $\alpha : (\alpha - 1)$ first-order resonance, this becomes

$$\mathcal{P}_{\text{syn}} \approx \mathcal{P}_{\text{tran}}\alpha\left[1 - (\alpha - 1)\Delta\right] \tag{5.62}$$

in the limit that $|\Delta| \ll 1$. Suppose that two objects with coplanar orbits are in conjunction at a time $t = 0$. At a time $t = \mathcal{P}_{\text{syn}}$, they are in conjunction again. If the objects are close to a first-order resonance, we can use Equation 5.62 to compute that between the two conjunctions, the line of conjunction rotates through an angle

$$\delta\theta_{\text{conj}} \approx -2\pi\alpha(\alpha - 1)\Delta. \tag{5.63}$$

After a time known as the **great inequality** timescale,[5]

$$\mathcal{P}_{\text{GI}} \approx \frac{2\pi}{|\delta\theta_{\text{conj}}|}\mathcal{P}_{\text{syn}} \approx \mathcal{P}_{\text{syn}}\frac{1}{\alpha(\alpha - 1)|\Delta|}, \tag{5.64}$$

the line of conjunction will return to its original orientation. For Kepler-88 b and Kepler-88 c, close to a $2:1$ resonance, we find a synodic period $\mathcal{P}_{\text{syn}} = 21.42\,\text{d}$ and a great inequality timescale $\mathcal{P}_{\text{GI}} \approx 25\mathcal{P}_{\text{syn}} \approx 540\,\text{d}$. Not by coincidence, the great inequality timescale is comparable to the period of the transit timing variations for Kepler-88 b, shown in Figure 5.8(b). The orbits of Kepler-88 b and c are moderately eccentric, with $e \sim 0.06$ for each orbit; thus, the distance between the exoplanets is not the same at every conjunction. If a particular conjunction occurs when the distance between the exoplanets is relatively small, then the 25th conjunction afterward will have a similar configuration and a similarly strong perturbation to the transiting exoplanet.

Being close to a first-order resonance, with $|\Delta| \ll 1$, will thus enhance the transit timing variations, given $\delta t \propto 1/|\Delta|$. However, some patience is required, since the timescale for variation of the TTVs is a relatively lengthy

$$\mathcal{P}_{\text{GI}} \sim \frac{\mathcal{P}_{\text{tran}}}{(\alpha - 1)|\Delta|}. \tag{5.65}$$

However, as sometimes happens in astronomy, patience pays off.

---

[5] This is named after the historical *Grande Inégalité* between Jupiter and Saturn. These two planets are near a $5:2$ resonance, with $\Delta = -0.0067$. This near-resonance causes the orbital periods of both planets to vary on a timescale much longer than their synodic period $\mathcal{P}_{\text{syn}} = 19.86\,\text{yr}$.

Exercises

5.1   A planet with radius R_2 orbits a star with radius R_1. The semimajor axis of the planet's orbit relative to the star is a. A distant observer (at $d \gg a$) is looking toward the star–planet system from a random location.

(a) Suppose the planet's orbit is circular ($e = 0$). Integrating over solid angle, show that the probability p_0 of the randomly positioned observer being able to see a transit (grazing or full) during the course of the planet's orbit is

$$p_0 = \frac{R_1 + R_2}{a}. \qquad (5.66)$$

(b) Now suppose the planet's orbit is eccentric ($0 < e < 1$). Integrating over solid angle, show that the probability p_e of the randomly positioned observer being able to see a transit during the course of the planet's orbit is

$$p_e = \frac{R_1 + R_2}{a(1 - e^2)}. \qquad (5.67)$$

5.2   Transit timing variations can be used to find "exomoons" (satellites of exoplanets). Suppose that a parent star of mass m_1 has a planet of mass $m_2 < m_1$; the planet in turn has a moon of mass $m_3 < m_2$. The moon is on a circular orbit relative to the planet, with radius a_{moon}. The barycenter of the planet–moon system is on a circular orbit relative to the parent star, with orbital radius $a_{plan} \gg a_{moon}$. The orbits are coplanar.

(a) What is the radius a of the planet's orbit relative to the planet–moon barycenter? What is the value of a, in kilometers, for the system consisting of the Earth and Moon?

(b) A distant observer detects transits of the planet across the parent star. Let $\delta t = \langle (O - C)^2 \rangle^{1/2}$ be a typical transit timing variation resulting from the moon's gravitational influence. Write an expression for $\delta t / \mathcal{P}$, where $\mathcal{P}$ is the orbital period of the planet–moon system around the parent star. (An approximate normalization is okay.)

(c) An alien astronomer near K2-65 b detects the Earth transiting the Sun. What value of δt (translating to human units such as "seconds" or "minutes") will the alien astronomer find as a result of the Moon's influence? Is this larger or smaller than δt from Jupiter's influence?

5.3   In the range $1 < \mathcal{P}_{pert}/\mathcal{P}_{tran} < 2$, what pair of orbits has the largest value of $|\Delta|$ for the nearest first-order resonance? What is the value of $\mathcal{P}_{pert}/\mathcal{P}_{tran}$ for this pair of orbits? What is the value of Δ for the two first-order resonances this pair of orbits lies between?

6

Many-Body Systems

A hundred thousand million Stars make one Galaxy;
a hundred thousand million Galaxies make one Universe.
The figures may not be very trustworthy,
but I think they give a correct impression.

Arthur Eddington (1882–1944)
The Expanding Universe, Chapter 1 [1933]

The techniques of celestial dynamics are useful within the solar system and other planetary systems. However, techniques that are useful in a system containing a few mutually gravitating objects are not as useful in a system containing a hundred thousand million objects. Even if it were practical to trace the orbits of all $\sim 3 \times 10^{11}$ stars in the Milky Way Galaxy, an exact knowledge of the position $\vec{r}$ and velocity $\vec{v}$ of every star would be too much information for most purposes.

The techniques of **stellar dynamics** are more useful than those of celestial dynamics when we study large stellar systems such as globular clusters and galaxies. Under most circumstances, we can treat stars as point masses attracting each other with a Newtonian gravitational force. Since stars are much bigger than their Schwarzschild radii,[1] tidal distortions and physical impacts between stars usually become important before general relativity needs to enter our arsenal.

When is it valid to assume that stars are point masses? As a typical stellar radius, let's take the Sun's radius:

$$1\,R_\odot = 696\,000\,\text{km} = 2.25 \times 10^{-8}\,\text{pc}. \tag{6.1}$$

If the average radius of two stars is $R_\star$, they will undergo a physical impact if their centers come within a distance $2R_\star$ of each other. Thus, the cross section σ for impacts is

---

[1] Things would be different in a very dense cluster of neutron stars, but we are looking at ordinary stars at ordinary number densities.

$$\sigma_{\text{imp}} = 4\pi R_\star^2 = 6.4 \times 10^{-15} \, \text{pc}^2 \left(\frac{R_\star}{1 \, \text{R}_\odot}\right)^2. \tag{6.2}$$

The mean free path λ of a star between physical impacts with another star is then

$$\lambda_{\text{imp}} = \frac{1}{n_\star \sigma_{\text{imp}}} = 7.8 \times 10^{14} \, \text{pc} \left(\frac{n_\star}{0.2 \, \text{pc}^{-3}}\right)^{-1} \left(\frac{R_\star}{1 \, \text{R}_\odot}\right)^{-2}, \tag{6.3}$$

scaling to the number density of stars in the solar neighborhood, $n_\star \sim 0.2 \, \text{pc}^{-3}$. Even by cosmological standards, λ_{imp} is a long distance: in the solar neighborhood, it's about 200 000 times the Hubble distance. The velocity dispersion of stars in the solar neighborhood is $v \sim 30 \, \text{km s}^{-1} \sim 3 \times 10^4 \, \text{pc} \, \text{Gyr}^{-1}$. This yields a typical time between impacts of

$$t_{\text{imp}} = \frac{\lambda_{\text{imp}}}{v} \tag{6.4}$$

$$= 2.6 \times 10^{10} \, \text{Gyr} \left(\frac{v}{30 \, \text{km s}^{-1}}\right)^{-1} \left(\frac{n_\star}{0.2 \, \text{pc}^{-3}}\right)^{-1} \left(\frac{R_\star}{1 \, \text{R}_\odot}\right)^{-2}.$$

Even by cosmological standards, t_{imp} is a very long time: in the solar neighborhood, it's about two billion times the Hubble time. The probability of the Sun physically colliding with another star before it swells into a red giant is roughly one in five billion.[2]

Usually, the motion of a single star in a galaxy can be approximated as that of a point mass moving through a smooth gravitational potential resulting from all the other stars in the galaxy. However, from time to time, a star will undergo a close gravitational encounter with another star.

We define a **close** gravitational encounter as an interaction that deflects the two stars' trajectory through an angle $\Delta\theta \geq 90°$. What initial conditions will produce such a large deflection? For concreteness, consider two stars of mass m_1 and m_2; the standard gravitational parameter of this two-body system is $\mu = G(m_1 + m_2)$. The two stars are not gravitationally bound to each other; that is, they start at a very large separation from each other, with a non-zero relative speed v and thus a positive specific energy $\epsilon = v^2/2$. The **impact parameter** of the two stars is b; that is, in the absence of gravity, the two stars (moving in straight lines at constant speed) would pass at a minimum distance b. If $b > 0$, the system has a non-zero orbital angular momentum, with $j = vb$. The gravitational attraction between the stars means that their relative motion follows a hyperbolic orbit with $e > 1$ and $a < 0$. For the deflection to have the value $\Delta\theta \geq 90°$, we require, from Equation 1.42, an eccentricity

$$e \leq e_c = \frac{1}{\sin(45°)} = \sqrt{2}. \tag{6.5}$$

---

[2] These odds will be somewhat lowered by tidal effects that increase the effective cross section; however, an impact between the Sun and another star is still highly improbable.

The eccentricity e of a hyperbolic orbit can be determined from the values of μ, v, and b. Since $\epsilon = -\mu/(2a) = v^2/2$, we can write

$$a = -\frac{\mu}{v^2}. \tag{6.6}$$

We also know, using Equation 1.43, that a hyperbolic orbit must have

$$a(1 - e^2) = \frac{j^2}{\mu} = \frac{b^2 v^2}{\mu}. \tag{6.7}$$

Combining Equations 6.6 and 6.7, we find that the eccentricity of a hyperbolic orbit is given by the relation

$$e^2 = 1 + \frac{v^4}{\mu^2}b^2. \tag{6.8}$$

Therefore, having $e^2 \le e_c^2 = 2$ requires an impact parameter

$$b \le b_{\mathrm{cge}} \equiv \frac{\mu}{v^2}. \tag{6.9}$$

Scaling to the velocity dispersion of stars in the solar neighborhood, and using $m_\star = (m_1 + m_2)/2$ as the mean stellar mass, we find the critical impact parameter is

$$b_{\mathrm{cge}} = \frac{2Gm_\star}{v^2} = 1.97\,\mathrm{au}\left(\frac{m_\star}{1\,M_\odot}\right)\left(\frac{v}{30\,\mathrm{km\,s^{-1}}}\right)^{-2}. \tag{6.10}$$

The cross section for a close gravitational encounter is

$$\sigma_{\mathrm{cge}} = \pi b_{\mathrm{cge}}^2 = 4\pi\frac{(Gm_\star)^2}{v^4}. \tag{6.11}$$

This differs from the cross section for physical impacts, $\sigma_{\mathrm{imp}} \approx \pi(2R_\star)^2$, by a factor

$$\frac{\sigma_{\mathrm{cge}}}{\sigma_{\mathrm{imp}}} \approx \left(\frac{Gm_\star}{R_\star}\right)^2\frac{1}{v^4} \sim \left(\frac{v_{\mathrm{esc},\star}}{v}\right)^4, \tag{6.12}$$

where $v_{\mathrm{esc},\star}$ is the escape speed from the surface of a star with mass $m_\star$. For Sun-like stars, $v_{\mathrm{esc},\star} \sim 600\,\mathrm{km\,s^{-1}}$; in the solar neighborhood, where the velocity dispersion of the stellar population is $v \sim 30\,\mathrm{km\,s^{-1}}$, the cross section for close gravitational encounters is larger than that for physical impacts by a factor of $\sim 200\,000$.

The typical time between close gravitational encounters for a given star is

$$t_{\mathrm{cge}} \sim \frac{1}{vn_\star\sigma_c} \sim \frac{v^3}{4\pi n_\star(Gm_\star)^2} \tag{6.13}$$

$$\sim 6 \times 10^5\,\mathrm{Gyr}\left(\frac{v}{30\,\mathrm{km\,s^{-1}}}\right)^3\left(\frac{n_\star}{0.2\,\mathrm{pc^{-3}}}\right)^{-1}\left(\frac{m_\star}{1\,M_\odot}\right)^{-2}.$$

The time between close gravitational encounters, t_{cge}, is also known as the two-body **relaxation time** t_{rel} for the stellar system in question. In the solar neighborhood, the relaxation time is $t_{rel} \sim 10^6 \, \text{Gyr}$; this is much larger than the age of the Milky Way Galaxy, $t_{sys} \sim 10 \, \text{Gyr}$. However, the central regions of globular clusters can have stellar densities as high as $n_\star \sim 1000 \, \text{pc}^{-3}$, and relatively small velocity dispersions of $v \sim 10 \, \text{km s}^{-1}$. Equation 6.13 then tells us that the relaxation time at the center of a globular cluster is $t_{rel} \sim 4 \, \text{Gyr}$.

In stellar dynamics, it is useful to make a distinction between stellar systems that are younger than their relaxation time and those that are older than their relaxation time. A stellar system with $t_{sys} < t_{rel}$ is known as a **collisionless system**. A system with $t_{sys} > t_{rel}$ is known as a **collisional system**. (This is not an ideal choice of names, since "collision," in this case, doesn't mean a physical impact, as it does in most contexts. Instead, it means a close gravitational encounter.) We will start our study of stellar dynamics by looking at collisionless systems, in which the occasional random "kicks" caused by close gravitational encounters can be ignored.

6.1 Virial Theorem

One of the most used (and abused) theorems in stellar dynamics is the **virial theorem**. The word "virial," seldom heard outside the context of the virial theorem, was coined by Rudolf Clausius in 1870; it is based on the Latin word *vis* (with plural form *vires*), meaning force or vigor.[3] The virial theorem is a very powerful tool; as such, it should be used with care, after reading the instructions and warning labels. In brief, the virial theorem relates the gravitational potential energy of a stellar system to the total kinetic energy of its orbiting stars; one use of the virial theorem is to estimate the mass of a stellar system such as a galaxy.

Suppose that a stellar system contains N point masses; let's call them "stars" for convenience. Each star has a mass m^α, where $\alpha = 1, 2, \ldots, N$. The total mass of the system is

$$M = \sum_\alpha m^\alpha. \tag{6.14}$$

We choose a coordinate system in which the barycenter of the system is at the origin. The position of each star in this barycentric coordinate system is $\vec{r}^\alpha$; its velocity is

$$\vec{v}^\alpha = \frac{d\vec{r}^\alpha}{dt}, \tag{6.15}$$

and its acceleration is

---

[3] Thus, "virial" is not cognate with "viral" (from the Latin *virus*, meaning "venom") or with "virile" (from the Latin *vir*, meaning "man").

$$\frac{d^2\vec{r}^\alpha}{dt^2} = \frac{\vec{F}^\alpha}{m^\alpha} = G\sum_{\beta\neq\alpha} m^\beta \frac{\vec{r}^\alpha - \vec{r}^\beta}{|\vec{r}^\alpha - \vec{r}^\beta|^3}. \tag{6.16}$$

Equation 6.16 assumes that the stellar system is **self-gravitating**. That is, the only force on each star is the net gravitational attraction from the other $N-1$ stars in the system. There are no significant external gravitational forces, and no significant non-gravitational forces.

Now it is time for some definitions. For convenience, we can write the position of each star in Cartesian coordinates as $\vec{r}^\alpha = (r_1^\alpha, r_2^\alpha, r_3^\alpha)$. The **moment of inertia tensor** of the stellar system is

$$I_{ij} = \sum_{\alpha=1}^{N} m^\alpha r_i^\alpha r_j^\alpha. \tag{6.17}$$

Given this definition, the moment of inertia tensor is a symmetric 3×3 tensor, with $I_{ij} = I_{ji}$. We may also define a moment of inertia **scalar**, which is simply the trace of the moment of inertia tensor:

$$I = \sum_{j=1}^{3} I_{jj} = \sum_\alpha m^\alpha |\vec{r}^\alpha|^2. \tag{6.18}$$

Thus, we can think of

$$\frac{I}{M} = \langle r^2 \rangle \tag{6.19}$$

as being the mass-weighted mean square distance of the stars from the barycenter.[4]

The **kinetic energy tensor** of the stellar system is

$$K_{ij} = \frac{1}{2}\sum_\alpha m^\alpha v_i^\alpha v_j^\alpha. \tag{6.20}$$

The kinetic energy tensor, like the moment of inertia tensor, is a symmetric 3×3 tensor. The kinetic energy **scalar** is

$$K = \sum_{j=1}^{3} K_{jj} = \frac{1}{2}\sum_\alpha m^\alpha |\vec{v}^\alpha|^2. \tag{6.21}$$

The kinetic energy scalar tells us how much energy is associated with the motion of stars in the barycentric frame. We can also find how much energy is associated with the gravitational forces between the stars. For a single pair of stars (call them α and β), the gravitational potential energy is

$$W^{\alpha\beta} = -\frac{Gm^\alpha m^\beta}{|\vec{r}^\alpha - \vec{r}^\beta|}. \tag{6.22}$$

---

[4] The scalar moment of inertia I that enters into the virial theorem differs from the more frequently used *rotational* moment of inertia $I_{\rm rot}$, which depends on the mean square distance of mass elements from an axis of rotation.

Generalizing to a system of N stars, we can define a **potential energy tensor**

$$W_{ij} = -\frac{1}{2}G\sum_{\alpha}\sum_{\beta\neq\alpha}m^{\alpha}m^{\beta}\frac{(r_i^{\alpha}-r_i^{\beta})(r_j^{\alpha}-r_j^{\beta})}{|(\vec{r}^{\alpha}-\vec{r}^{\beta})|^3}. \tag{6.23}$$

Once again, we have constructed a symmetric 3×3 tensor. The potential energy **scalar** is then

$$W = \sum_{j=1}^{3}W_{jj} = -\frac{1}{2}G\sum_{\alpha}\sum_{\beta\neq\alpha}\frac{m^{\alpha}m^{\beta}}{|\vec{r}^{\alpha}-\vec{r}^{\beta}|}. \tag{6.24}$$

(The factor of $1/2$ in Equations 6.23 and 6.24 ensures that each pair of stars is counted only once when adding up the total gravitational potential energy.)

With these definitions set up, we can plunge into the physics of the virial theorem. Start by taking the second derivative of the moment of inertia tensor (Equation 6.17) with respect to time:

$$\frac{d^2I_{ij}}{dt^2} = \sum_{\alpha}m^{\alpha}\left(\frac{d^2r_i^{\alpha}}{dt^2}r_j^{\alpha} + 2\frac{dr_i^{\alpha}}{dt}\frac{dr_j^{\alpha}}{dt} + r_i^{\alpha}\frac{d^2r_j^{\alpha}}{dt^2}\right). \tag{6.25}$$

Using the definition of the kinetic energy tensor (Equation 6.20), we can simplify this as

$$\frac{d^2I_{ij}}{dt^2} = 4K_{ij} + \sum_{\alpha}m^{\alpha}\left(\frac{d^2r_i^{\alpha}}{dt^2}r_j^{\alpha} + r_i^{\alpha}\frac{d^2r_j^{\alpha}}{dt^2}\right). \tag{6.26}$$

Using Newton's second law of motion, we can rewrite the above equation as

$$\frac{d^2I_{ij}}{dt^2} = 4K_{ij} + \sum_{c}\left(F_i^{\alpha}r_j^{\alpha} + r_i^{\alpha}F_j^{\alpha}\right), \tag{6.27}$$

where F_i^{α} is the net force acting on star α in the r_i direction.

Equation 6.27 is true no matter what forces are acting on the stars. In 1870, Clausius derived the virial theorem for what he called "any system whatever of material points in stationary motion," explaining that "stationary motion" implied motion within a finite volume with an upper limit on particle velocities. In 1911, Henri Poincaré applied the virial theorem to a system of planetesimals destined to become the solar system; in 1916, Arthur Eddington extended the virial theorem to the study of star clusters. In a self-gravitating system such as an isolated star cluster or a galaxy, the forces are given by Equation 6.16, and Equation 6.27 becomes

$$\frac{d^2I_{ij}}{dt^2} = 4K_{ij} + G\sum_{\alpha}\sum_{\beta\neq\alpha}m^{\alpha}m^{\beta}\left[\frac{r_j^{\alpha}(r_i^{\alpha}-r_i^{\beta}) + r_i^{\alpha}(r_j^{\alpha}-r_j^{\beta})}{|\vec{r}^{\alpha}-\vec{r}^{\beta}|^3}\right]. \tag{6.28}$$

Since α and β are dummy indices that we are summing over, we can replace α with β and β with α. This changes Equation 6.28 into

$$\frac{d^2 I_{ij}}{dt^2} = 4K_{ij} + G \sum_{\beta} \sum_{\alpha \neq \beta} m^\beta m^\alpha \left[\frac{r_j^\beta (r_i^\beta - r_i^\alpha) + r_i^\beta (r_j^\beta - r_j^\alpha)}{|\vec{r}^\beta - \vec{r}^\alpha|^3} \right]. \tag{6.29}$$

Adding together Equations 6.28 and 6.29, we find

$$2\frac{d^2 I_{ij}}{dt^2} = 8K_{ij} + 2G \sum_{\alpha, \beta \neq \alpha} m^\alpha m^\beta \left[\frac{r_i^\alpha (r_j^\alpha - r_j^\beta) - r_i^\beta (r_j^\alpha - r_j^\beta)}{|\vec{r}^\beta - \vec{r}^\alpha|^3} \right]. \tag{6.30}$$

Looking back at the definition of the potential energy tensor in Equation 6.23, we realize we can write this compactly (after dividing by four) as

$$\frac{1}{2}\frac{d^2 I_{ij}}{dt^2} = 2K_{ij} + W_{ij}. \tag{6.31}$$

Equation 6.31 is the usual form of the **tensor virial theorem**. Taking the trace of the tensor virial theorem, we find the **scalar virial theorem**:

$$\frac{1}{2}\frac{d^2 I}{dt^2} = 2K + W. \tag{6.32}$$

If we assume that the moment of inertia scalar I is constant with time (that is, if the stellar system isn't expanding or contracting), we end with the **steady-state scalar virial theorem**:

$$K = -\frac{1}{2}W. \tag{6.33}$$

Eddington expressed this result as: "In any Star Cluster in a steady state, the Internal Kinetic Energy is one-half the exhaustion of Potential Energy."

Equation 6.33 is appealingly simple, but we must review the assumptions that went into it. Is the stellar system in question actually in a steady state? Is it self-gravitating? Applying the steady-state scalar virial theorem to a system that is pulsating inward or outward, or a system being ripped apart by tides, will give misleading results.

The kinetic energy scalar has a positive value, $K > 0$; the potential energy scalar has a negative value, $W < 0$. In a steady-state bound system, the total energy is, using Equation 6.33,

$$E \equiv K + W = -K = \frac{1}{2}W < 0. \tag{6.34}$$

Thinking of the stellar system as a confined gas of stars, we can assign a "temperature" T by using the relation

$$K = \frac{3}{2}NkT. \tag{6.35}$$

The total energy of the stellar system is then

$$E = -\frac{3}{2} NkT \tag{6.36}$$

and its heat capacity is

$$C \equiv \frac{dE}{dT} = -\frac{3}{2} Nk. \tag{6.37}$$

Thus, the heat capacity of a self-gravitating system is negative; if energy is added, the temperature (or equivalently, the mean kinetic energy of the stars) decreases. At the same time, the system becomes less strongly bound.

A common use of the virial theorem is to estimate the mass of a stellar system. The total kinetic energy of the system can be written as

$$K = \frac{1}{2} M \langle v^2 \rangle, \tag{6.38}$$

where

$$M = \sum_\alpha m^\alpha \tag{6.39}$$

is the total mass of the system, and

$$\langle v^2 \rangle = \frac{1}{M} \sum_\alpha m^\alpha |\vec{v}^\alpha|^2 \tag{6.40}$$

is the mass-weighted mean square velocity of the individual stars. The total potential energy of the system can be written as

$$W = -\alpha_w \frac{GM^2}{r_h}, \tag{6.41}$$

where r_h is the **half-mass radius** of the system; that is, it is the radius of a sphere centered on the system's barycenter that contains half the total mass M of the system. The parameter α_w is a dimensionless factor that depends on the density profile of the stellar system. For a uniform-density sphere of radius R, the half-mass radius is $r_h = R/2^{1/3}$, and the potential energy is

$$W = -\frac{3}{5} \frac{GM^2}{R} = -\frac{3}{5(2^{1/3})} \frac{GM^2}{r_h} \approx -0.476 \frac{GM^2}{r_h}. \tag{6.42}$$

For more realistic models of stellar systems, it is usually found that $\alpha_w \approx 0.5$. Since the steady-state scalar virial theorem tells us that $K = -W/2$, we can substitute from Equations 6.38 and 6.41 to find

$$\frac{1}{2} M \langle v^2 \rangle = \frac{\alpha_w}{2} \frac{GM^2}{r_h}. \tag{6.43}$$

Solving for M, we find the mass of a system that obeys the steady state virial theorem:

$$M = \frac{\langle v^2 \rangle r_h}{\alpha_w G}. \tag{6.44}$$

Equation 6.44 is useful, with the proper precautions, for estimating the mass of a stellar system.

What precautions are appropriate? First, we don't generally know the mass-weighted three-dimensional mean square velocity $\langle v^2 \rangle$. Instead, we usually know, from the spectrum of the stellar system, a luminosity-weighted line-of-sight velocity dispersion σ_{los}. The dark matter provides none of the photons from which we determine σ_{los}; if we are observing at visible wavelengths, most of them come from a relatively small number of highly luminous stars. Second, we don't generally know the three-dimensional half-mass radius r_h. Instead, we usually deproject the surface brightness of the system to estimate a three-dimensional half-light radius r_{hl}; this is not necessarily the same as the half-mass radius. Thus, the usual virial mass estimate,

$$M \approx 3 \frac{\sigma_{los}^2 r_{hl}}{\alpha_w G} \approx 6 \frac{\sigma_{los}^2 r_{hl}}{G}, \tag{6.45}$$

should be used only as an estimate.

6.2 Collisionless Boltzmann Equation

If a system contains a sufficiently large number of stars, the **distribution function** in phase space becomes a useful tool. Choose a location $\vec{r}$ within a large stellar system such as a galaxy. Draw a volume element d^3r around that point, small compared to the total volume of the stellar system, but large enough to contain a great many stars. Then choose a velocity $\vec{v}$ and draw a volume element d^3v around that velocity point. The distribution function $f(\vec{r}, \vec{v}, t)$ is then defined so that $f(\vec{r}, \vec{v}, t) d^3r d^3v$ is the number of stars that are (at time t) in the volume element d^3r centered on $\vec{r}$ and in the velocity volume element d^3v centered on $\vec{v}$. The distribution function defined in this way is thus a number density of stars in six-dimensional phase space. We can also define a **mass** distribution function $f_m(\vec{r}, \vec{v}, t)$, which gives the mass density of stars in phase space. If stars of different mass have the same distribution in position and velocity, then $f_m = m_\star f$, where $m_\star$ is the mean stellar mass; however, there are often situations where high-mass stars have a different spatial distribution or velocity distribution than low-mass stars. In a similar way, we can define a **luminosity** distribution function $f_L(\vec{r}, \vec{v}, t)$, which gives the luminosity density of stars in phase space.

Let's assume that stars are neither created nor destroyed. Although astronomers who study star formation and stellar mergers may feel slighted by this assumption, it is a useful approximation in the context of stellar dynamics. Let's also assume that the stellar system is collisionless, in the sense that close gravitational encounters between stars don't cause abrupt kicks to any star's velocity. In this case, there must be a **continuity equation** in phase space. Expressed in Cartesian coordinates, the continuity equation is

$$\frac{\partial f}{\partial t} + \sum_{i=1}^{3} \frac{\partial}{\partial r_i}\left(f\frac{dr_i}{dt}\right) + \sum_{i=1}^{3} \frac{\partial}{\partial v_i}\left(f\frac{dv_i}{dt}\right) = 0. \tag{6.46}$$

That is, if the distribution function f is changing at any point in six-dimensional phase space, it is because there is a divergence or convergence of the flux of stars at that point. However, we can write $dr_i/dt = v_i$. Also, if the only relevant force is gravity, we can write

$$\frac{dv_i}{dt} = -\frac{\partial \Phi}{\partial r_i}, \tag{6.47}$$

where $\Phi(\vec{r}, t)$ is the local gravitational potential. With the appropriate substitutions, Equation 6.46 becomes

$$\frac{\partial f}{\partial t} + \sum_i \frac{\partial}{\partial r_i}(fv_i) - \sum_i \frac{\partial}{\partial v_i}\left(f\frac{\partial \Phi}{\partial r_i}\right) = 0. \tag{6.48}$$

Since $\vec{r}$ and $\vec{v}$ are independent variables, and since the potential Φ is independent of $\vec{v}$, Equation 6.48 can be written as

$$\frac{\partial f}{\partial t} + \sum_i v_i\frac{\partial f}{\partial r_i} - \sum_i \frac{\partial \Phi}{\partial r_i}\frac{\partial f}{\partial v_i} = 0. \tag{6.49}$$

Equation 6.49 is called the **collisionless Boltzmann equation**. Ludwig Boltzmann first derived a similar equation in 1872, in the context of fluid dynamics. However, it can be applied to point masses interacting via gravity as well as to molecules and atoms interacting at short ranges.

In vector notation, the collisionless Boltzmann equation is

$$\frac{\partial f}{\partial t} + \vec{v} \cdot \vec{\nabla}f - \vec{\nabla}\Phi \cdot \frac{\partial f}{\partial \vec{v}} = 0. \tag{6.50}$$

The collisionless Boltzmann equation has been described as the fundamental equation of stellar dynamics; it has also been described as the fundamental equation of fluid dynamics. It is undeniably a Very Important Equation. However, in the context of stellar dynamics, it is not a very practical equation. Typically, we don't have enough information about a stellar system to determine its distribution function $f(\vec{r}, \vec{v})$ at a given time t. Consider, for instance, a rich globular cluster with 10^6 stars. If we divide each of the axes of phase space into 10 bins, that's 10^6 volume elements $d^3r\,d^3v$. With the number of volume elements equal to the number of stars, we just don't have enough data to determine f at the proposed resolution. The basic problem is that stellar systems don't contain enough stars; the number of stars in our galaxy is less than the number of molecules in a cube of air a hairsbreadth across (assuming a hair is $\sim 70\ \mu m$ in diameter).

Instead of using the distribution function f, in stellar dynamics is it usually more useful to take the moments of the distribution function, averaged over velocity space. For instance, integrating f over all velocities yields the **number density** of stars in space:

$$n_\star(\vec{r}, t) = \int f(\vec{r}, \vec{v}, t)\, d^3v. \tag{6.51}$$

Similarly, the **mass density** of stars in space is

$$\rho_\star(\vec{r}, t) = \int f_m(\vec{r}, \vec{v}, t)\, d^3v \tag{6.52}$$

and the **luminosity density** is

$$\Psi_\star(\vec{r}, t) = \int f_L(\vec{r}, \vec{v}, t)\, d^3v. \tag{6.53}$$

If we regard a stellar system as being a collection of N point masses, its gravitational potential is

$$\Phi(\vec{r}) = -\sum_{\alpha=1}^{N} \frac{Gm^\alpha}{|\vec{r}^{\,\alpha} - \vec{r}|}. \tag{6.54}$$

The deep potential wells around each point mass are the sources of the abrupt kicks that a star receives during a close gravitational encounter. However, in a collisionless system, it is safe to approximate the density as a smooth function. In this approximation, the gravitational potential is given by the integral

$$\Phi(\vec{r}) = -G \int \frac{\rho(\vec{x})\, d^3x}{|\vec{x} - \vec{r}|}, \tag{6.55}$$

where $\rho(\vec{r})$ is the *total* mass density, which in some cases may include contributions from dark matter or interstellar gas; thus, we expect $\rho \geq \rho_\star$.

Taking the Laplacian on both sides of Equation 6.55, we find Poisson's equation:

$$\nabla^2 \Phi = 4\pi G \rho(\vec{r}). \tag{6.56}$$

The use of a smooth density ρ also changes the form of the potential energy scalar W. For a system of N point masses, we found (from Equation 6.24)

$$W = -\frac{1}{2} G \sum_\alpha \sum_{\beta \neq \alpha} \frac{m^\alpha m^\beta}{|\vec{r}^{\,\alpha} - \vec{r}^{\,\beta}|} = \frac{1}{2} \sum_\alpha m^\alpha \Phi(\vec{r}^{\,\alpha}). \tag{6.57}$$

If we go from a sum to an integral, the potential energy becomes

$$W = \frac{1}{2} \int \rho(\vec{r}) \Phi(\vec{r}) \, d^3r. \tag{6.58}$$

Integrating the distribution function $f(\vec{r}, \vec{v})$ over all velocities yields the number density $n_\star(\vec{r})$ of stars. If we multiply the distribution function by $\vec{v}$ before integrating, we find the **bulk velocity** $\vec{u}$ of the stars in the volume element d^3r at location $\vec{r}$:

$$\vec{u}(\vec{r}, t) = \frac{1}{n_\star} \int \vec{v} f(\vec{r}, \vec{v}, t) \, d^3v. \tag{6.59}$$

Similarly, we can find the bulk velocity weighted by the mass of stars, or by the luminosity of stars.

Going to the next higher order moment, the mean square velocity of the stars at location $\vec{r}$ and time t is

$$\langle v_i v_j \rangle = \frac{1}{n_\star} \int v_i v_j f(\vec{r}, \vec{v}, t) \, d^3v. \tag{6.60}$$

From the mean square velocity and the bulk velocity, we can compute the **velocity dispersion**:

$$\sigma_{ij}^2 = \langle v_i v_j \rangle - u_i u_j. \tag{6.61}$$

The velocity dispersion is a symmetric tensor, with $\sigma_{ij} = \sigma_{ji}$. This means that at any location $\vec{r}$, there exists a set of Cartesian axes for which the velocity dispersion tensor is diagonal, with $\sigma_{ij} = 0$ when $i \neq j$. The three axes of the diagonalized velocity dispersion define a **velocity ellipsoid**. In a collisionless stellar system, the three axes of the velocity ellipsoid are not necessarily equal in length.

It is possible, of course, to take higher moments of the distribution function to find $\langle v_i v_j v_k \rangle$, $\langle v_i v_j v_k v_\ell \rangle$, and so forth. In practice, however, the bulk velocity and the velocity dispersion are the most useful moments of the stellar velocity distribution.

6.3 Jeans Equations

Just as we can take moments of the dispersion function f, we can take moments of the collisionless Boltzmann function. For example, if we simply take the collisionless Boltzmann function (Equation 6.49) and integrate over all velocities, we find

$$\frac{\partial}{\partial t} \int f \, d^3v + \sum_i \frac{\partial}{\partial r_i} \int f v_i \, d^3v - \sum_i \frac{\partial \Phi}{\partial r_i} \int \frac{\partial f}{\partial v_i} \, d^3v = 0. \tag{6.62}$$

In Equation 6.62, the first integral is the number density $n_\star(\vec{r}, t)$. The second integral, from Equation 6.59, is $n_\star u_i$. The third integral, from the divergence theorem, is equal to zero if $f \to 0$ as $v \to \infty$. This should be the case: we do not expect any infinitely fast stars. Thus, Equation 6.62 can be written as

$$\frac{\partial n_\star}{\partial t} + \sum_i \frac{\partial}{\partial r_i}(n_\star u_i) = 0. \tag{6.63}$$

This is a **continuity equation**: it states that stars don't pop out of nowhere, or vanish into nothingness, or travel by instantaneous teleportation.

If we multiply the collisionless Boltzmann equation by a velocity component v_j, then integrate over all velocities, we find

$$\frac{\partial}{\partial t} \int v_j f \, d^3v + \sum_i \frac{\partial}{\partial r_i} \int f v_i v_j \, d^3v - \sum_i \frac{\partial \Phi}{\partial r_i} \int \frac{\partial f}{\partial v_i} v_j \, d^3v = 0. \tag{6.64}$$

In Equation 6.64, the first integral is $n_\star u_j$. The second integral, from Equation 6.60, is $n_\star \langle v_i v_j \rangle$. The third integral, after integrating by parts and applying the divergence theorem, becomes $-n_\star \delta_{ij}$, where δ_{ij} is the Kronecker delta, equal to one if $i = j$ and equal to zero otherwise. Thus, Equation 6.64 can be written as

$$\frac{\partial}{\partial t}\left(n_\star u_j\right) + \sum_i \frac{\partial}{\partial r_i}\left(n_\star \langle v_i v_j \rangle\right) = -n_\star \frac{\partial \Phi}{\partial r_j}. \tag{6.65}$$

This is a **momentum conservation equation**: the momentum density in a volume element changes if there is a convergence or divergence of the momentum flow, or if there is a net gravitational force acting. With the use of Equation 6.63 we can rewrite Equation 6.65 as an equation of motion:

$$\frac{\partial u_j}{\partial t} + \sum_i u_i \frac{\partial u_j}{\partial r_i} = -\frac{\partial \Phi}{\partial r_j} - \frac{1}{n_\star} \sum_i \frac{\partial}{\partial r_i}(n_\star \sigma_{ij}^2). \tag{6.66}$$

The left-hand side of Equation 6.66 represents the Lagrangian time derivative of u_j; that is, it is the time derivative as measured by an observer moving with the bulk velocity $\vec{u}$. The right-hand side contains a term from the gravitational potential and a term that is the equivalent of a pressure gradient. Thus, for a steady-state system, Equation 6.66 is analogous to the equation of hydrostatic equilibrium in fluid dynamics.

Equation 6.63 and the three versions of Equation 6.65 (for $j = 1, 2,$ and 3) are collectively known as the **Jeans equations**. However, the four Jeans equations are badly outnumbered by the unknowns. Suppose you know perfectly the stellar number density $n_\star$ and the total mass density ρ (which in turn tells you Φ). There still remain the three components of the bulk velocity $\vec{u}$ and the six independent components of the velocity dispersion tensor σ_{ij}. The Jeans equations can be solved only if simplifying assumptions are made.

For some isolated systems it is possible to adopt the steady-state approximation, in which $n_\star$, $\vec{u}$, and σ_{ij} are not explicitly dependent on time. In the steady-state approximation, it is not necessary for the bulk velocity $\vec{u}$ to be zero everywhere; however, it does have to represent a flow of stars that leaves $n_\star(\vec{r})$ unchanged with time. In the steady-state approximation, systematic expansion, contraction, or shear of a stellar system is not permitted.

One application of the steady-state Jeans equations is to axisymmetric stellar systems; a disk galaxy is nearly axisymmetric in many cases. In such a system, cylindrical coordinates (R, z, θ) are useful: R is the distance from the axis of rotational symmetry, z is the distance from an equatorial plane perpendicular to the axis of symmetry, and θ is the azimuthal angle. In a steady-state axisymmetric system, $u_R = u_z = 0$; however, non-zero u_θ is permitted. Since derivatives with respect to t and θ vanish in a steady-state axisymmetric system, the momentum continuity equations (Equation 6.65) can be written in cylindrical coordinates as

$$\frac{1}{R}\frac{\partial}{\partial R}\left(Rn_\star\sigma_{RR}^2\right) + \frac{\partial}{\partial z}\left(n_\star\sigma_{Rz}^2\right) - n_\star\frac{\sigma_{\theta\theta}^2 + u_\theta^2}{R} = -n_\star\frac{\partial\Phi}{\partial R}, \qquad (6.67)$$

$$\frac{1}{R^2}\frac{\partial}{\partial R}\left(R^2 n_\star\sigma_{R\theta}^2\right) + \frac{\partial}{\partial z}\left(n_\star\sigma_{z\theta}^2\right) = 0, \qquad (6.68)$$

$$\frac{1}{R}\frac{\partial}{\partial R}\left(Rn_\star\sigma_{Rz}^2\right) + \frac{\partial}{\partial z}\left(n_\star\sigma_{zz}^2\right) = -n_\star\frac{\partial\Phi}{\partial z}. \qquad (6.69)$$

Another application of the steady-state Jeans equations is to a spherically symmetric system; globular clusters are close to spherical symmetry, as are some elliptical galaxies. In such a system, spherical coordinates (r, θ, ϕ) are useful: r is the distance from the system's center, θ is the polar angle, and ϕ is the azimuthal angle. In steady state, $u_r = 0$; we also assume there is no net rotation of the system, so $u_\theta = u_\phi = 0$. With no rotation to break the spherical symmetry, our choice of "north pole" and "prime meridian" for measuring θ and ϕ is arbitrary; this means, for instance, that we expect the velocity dispersion to be isotropic on the θ–ϕ plane, with $\sigma_{\theta\theta} = \sigma_{\phi\phi} = \sigma_t$. In addition, spherical symmetry leads us to expect $\sigma_{r\theta} = \sigma_{r\phi} = \sigma_{\theta\phi} = 0$. With these simplifications, the only non-trivial momentum conservation equation (Equation 6.65) is the radial equation, which becomes

$$\frac{d}{dr}\left(n_\star\sigma_{rr}^2\right) + 2n_\star\frac{\sigma_{rr}^2 - \sigma_t^2}{r} = -n_\star\frac{d\Phi}{dr}. \qquad (6.70)$$

Given our assumptions, there is no requirement that the radial dispersion σ_{rr} be equal to the transverse dispersion σ_t. The anisotropy of the velocity dispersion can be usefully described by the **anisotropy parameter**

$$\beta(r) \equiv 1 - \frac{\sigma_{\theta\theta}^2 + \sigma_{\phi\phi}^2}{2\sigma_{rr}^2} = 1 - \frac{\sigma_t^2}{\sigma_{rr}^2}. \qquad (6.71)$$

A spherical stellar system with stars on nearly radial orbits will have $\beta \approx 1$, while a system with stars on purely circular orbits will have $\beta = -\infty$. In terms of the anisotropy parameter, the spherical Jeans equation (Equation 6.70) can be written as

$$\frac{d}{dr}\left(n_\star \sigma_{rr}^2\right) + 2\beta \frac{n_\star \sigma_{rr}^2}{r} = -n_\star \frac{d\Phi}{dr}. \tag{6.72}$$

In the limit of perfect isotropy ($\beta = 0$), this is equivalent to the equation of hydrostatic equilibrium for a sphere (familiar to those of you who have studied stellar or planetary interiors).

6.4 Spherical Systems

For an arbitrary mass distribution $\rho(\vec{r})$, computing the gravitational potential $\Phi(\vec{r})$ from Equation 6.55 is a task best done numerically. However, the computation for a spherically symmetric distribution is much simpler, thanks to Newton's two **shell theorems**. Expressed in modern language rather than the Latin of the *Principia Mathematica*, the shell theorems state:

- An object *inside* a spherically symmetric mass shell experiences no net gravitational force from that shell.
- An object *outside* a spherically symmetric mass shell experiences the same gravitational force it would feel if all the shell's matter were squeezed into a point at the shell's center.

Since any spherically symmetric matter distribution can be divided into concentric spherically symmetric shells, the shell theorems imply that an object at a distance r from the center of a spherical mass distribution feels the same force it would experience if all the mass $M(r)$ inside its location were concentrated as a point mass at the distribution's center.

Given the appealing simplicity of spherical stellar systems (due in large part to the shell theorems), it is tempting to approximate every galaxy as being spherical. In fact, the "spherical stellar system" approximation is often much better than the "spherical cow" approximation. In particular, the potential $\Phi(\vec{r})$ of a stellar system is more nearly spherical than its mass density $\rho(\vec{r})$, since Φ is found by smoothing ρ with a very broad $1/r$ kernel (Equation 6.55). As an example, Figure 6.1(b) shows the equipotential contours for the axisymmetric potential

$$\Phi(R, z) = \frac{1}{2}v_0^2 \ln(R^2 + z^2/q^2), \tag{6.73}$$

where $q = 0.8$. The equipotential contours are similar oblate spheroids whose axis ratio $c/a = q = 0.8$ is not much smaller than unity. However, Figure 6.1(a) shows the isodensity contours for the mass density that produces the potential

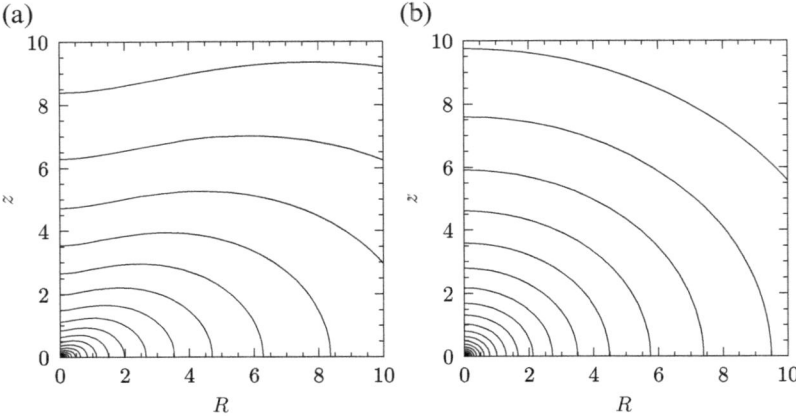

Figure 6.1 (a) Isodensity contours for the mass distribution of Equation 6.74, assuming $q = 0.8$; the contour spacing is $\Delta \log \rho = 0.25$. (b) Equipotential contours for the associated gravitational potential, given in Equation 6.73; the contour spacing is $\Delta \Phi = 0.25 v_0^2$.

in Equation 6.73. Solving Poisson's equation, we find that the required mass density is

$$\rho(R, z) = \frac{v_0^2}{4\pi G q^2} \frac{R^2 + z^2(2 - 1/q^2)}{(R^2 + z^2/q^2)^2}. \tag{6.74}$$

The isodensity contours, Figure 6.1(a), are more flattened than the equipotential contours, Figure 6.1(b), with an axis ratio

$$\frac{z}{R} = q^2(2 - 1/q^2)^{1/2} \approx 0.42. \tag{6.75}$$

Also notice from Equation 6.74 that flattening the potential to an axis ratio $q < 1/\sqrt{2} \approx 0.71$ would require a *negative* mass density along the z axis.[5]

For a perfectly spherical mass distribution $\rho(r)$, the corresponding gravitational potential (Equation 6.55) is

$$\Phi(r) = -4\pi G \left[\frac{1}{r} \int_0^r \rho(x) x^2 \, dx + \int_r^\infty \rho(x) x \, dx \right]. \tag{6.76}$$

The resulting gravitational acceleration at any point $\vec{r}$ is

$$\frac{d^2 \vec{r}}{dt^2} = -\frac{d\Phi}{dr} \hat{r} = -\frac{GM(r)}{r^2} \hat{r}, \tag{6.77}$$

where $\hat{r} = \vec{r}/r$ and

$$M(r) = 4\pi \int_0^r \rho(x) x^2 \, dx \tag{6.78}$$

is the mass interior to r.

[5] This would give Isaac Newton a very bad headache indeed.

Suppose we place an infinitesimal test mass in a spherical potential and let it move purely under the influence of gravity. To move on a circular orbit, it must have a speed equal to the **circular speed**

$$v_{\text{circ}}(r) = \left(\frac{GM(r)}{r} \right)^{1/2}. \tag{6.79}$$

To be on an unbound orbit, it must have a speed equal to or greater than the **escape speed**. The escape speed $v_{\text{esc}}(r)$ is found by setting the kinetic energy of the test mass equal to the gravitational potential energy it gains in going from r to ∞. Using the usual convention that $\Phi \to 0$ as $r \to \infty$, we can write

$$v_{\text{esc}}(r) = [-2\Phi(r)]^{1/2}. \tag{6.80}$$

Although the circular speed depends only on the amount of mass inside r, the escape speed also depends on how the mass is distributed outside r.

If a spherical density distribution contains a population of luminous test masses, all moving on circular orbits, the speed of the test masses provides a useful mass estimator. From Equation 6.79,

$$GM(r) = v_{\text{circ}}^2 r. \tag{6.81}$$

Stars in disk galaxies move on nearly circular orbits, and thus can be used to estimate the mass of the nearly spherical dark halo in which they orbit. For a spherical stellar system with no net rotation, we can use the radial Jeans equation (Equation 6.72) to construct a mass estimator. Using the relation $d\Phi/dr = -GM/r^2$, Equation 6.72 can be rewritten as

$$GM(r) = \sigma_{rr}^2 r \left[-\frac{d\ln(n_\star \sigma_{rr}^2)}{d\ln r} - 2\beta \right]. \tag{6.82}$$

The virial theorem gave us an estimate for the total mass of a system (Equation 6.44). The radial Jeans equation gives an estimate for the mass profile $M(r)$ of a spherical system. In both cases, however, the anisotropy of the velocity dispersion must be known for an accurate mass estimate.

6.4.1 Power-Law Density Profiles

As a simple example, consider a spherical stellar system with mass density $\rho \propto r^{-\eta}$ and $\eta \geq 0$. The central singularity, with $\rho \to \infty$ as $r \to 0$, is not necessarily a fatal flaw. As long as $\eta < 3$, the mass interior to r decreases with radius as $M \propto r^{3-\eta}$; thus at some finite radius, the enclosed mass will be comparable to the mass of a single star, and the assumption of smooth density and potential will break down. When $\eta < 3$, the enclosed mass becomes infinite as $r \to \infty$; we can avoid stellar systems with infinite mass by truncating the mass distribution at a finite radius r_t. Given $\eta < 3$, the density profile at $r \leq r_t$ is

$$\rho(r) = \frac{3 - \eta}{4\pi} \frac{M_t}{r_t^3} \left(\frac{r}{r_t}\right)^{-\eta} \tag{6.83}$$

and the mass profile is

$$M(r) = M_t \left(\frac{r}{r_t}\right)^{3-\eta}, \tag{6.84}$$

where M_t is the total mass of the system, with

$$M_t = \frac{4\pi}{3 - \eta} \rho(r_t) r_t^3. \tag{6.85}$$

For $\eta \neq 2$, the potential interior to the truncation radius r_t is

$$\Phi(r) = -\frac{GM_t}{r_t} \left(\frac{3 - \eta}{2 - \eta}\right) \left[1 - \frac{1}{3 - \eta}(\frac{r}{r_t})^{2-\eta}\right]. \tag{6.86}$$

Thanks to the second shell theorem, the escape speed from the boundary of the stellar system,

$$v_{esc}(r_t) = \left(\frac{2GM_t}{r_t}\right)^{1/2}, \tag{6.87}$$

depends only on M_t and r_t. However, the escape speed from the central region, where $r \ll r_t$, depends strongly on η. When $\eta < 2$, the escape speed at the center is finite, with

$$\frac{v_{esc}(0)}{v_{esc}(r_t)} = \left(\frac{3 - \eta}{2 - \eta}\right)^{1/2} > 1. \tag{6.88}$$

When $2 < \eta < 3$, the escape speed goes to infinity at the center of the system, with

$$\frac{v_{esc}(r \ll r_t)}{v_{esc}(r_t)} \approx \frac{1}{(\eta - 2)^{1/2}} \left(\frac{r}{r_t}\right)^{1-\eta/2}. \tag{6.89}$$

Using Equation 6.58, the potential energy of the spherical distribution is finite when $\eta < 5/2$, with

$$W = -\frac{3 - \eta}{5 - 2\eta} \frac{GM_t^2}{r_t}. \tag{6.90}$$

Thus, for a system of given mass M_t and radius r_t, a steeper density profile leads to a deeper potential well and a larger value of $|W|$. However, a more steeply declining density profile also leads to a smaller half-mass radius, with $r_h = 2^{-1/(3-\eta)} r_t$. Thus, if we write the potential energy in the form of Equation 6.41,

$$W = -\alpha_w \frac{GM_t^2}{r_h}, \tag{6.91}$$

we find that

$$\alpha_w(\eta) = 2^{-1/(3-\eta)} \frac{3-\eta}{5-2\eta}, \tag{6.92}$$

which lies in the narrow range $0.47 < \alpha_w \le 0.50$ for $0 \le \eta \le 2$.

Equation 6.82 tells us that there is not a unique solution for the velocity dispersion of stars within a known mass distribution $M(r)$. If we assume, for convenience, that the dispersion is isotropic, with $\beta = 0$ and $\sigma_{rr} = \sigma_t = \sigma$, then Equation 6.82 reduces to

$$\sigma(r)^2 = \frac{GM_t}{r_t} \left(\frac{r}{r_t}\right)^{2-\eta} \left[-\frac{d\ln(n_\star \sigma^2)}{d\ln r}\right]^{-1}, \tag{6.93}$$

assuming a power-law distribution of mass, with $\rho \propto r^{-\eta}$.

A particularly useful application of the power-law density profile is the **singular isothermal sphere**, for which σ is constant. For $\eta = 2$, Equation 6.93 has the solution

$$\sigma^2 = \frac{GM_t}{r_t} \left[-\frac{d\ln n_\star}{d\ln r}\right]^{-1}. \tag{6.94}$$

If the number density of stars traces the mass density, then $d\ln n_\star / d\ln r = -2$, and the distribution is an singular isothermal sphere; its velocity dispersion (analogous to a thermal speed) is

$$\sigma^2 = \frac{GM_t}{2r_t} = 2\pi G\rho(r_t)r_t^2. \tag{6.95}$$

Similarly, the speed of a test particle on a circular orbit is independent of r in a singular isothermal sphere, with (Equation 6.79)

$$v_{\text{circ}}^2 = \frac{GM_t}{r_t} = 4\pi G\rho(r_t)r_t^2 = 2\sigma^2. \tag{6.96}$$

One application of singular isothermal spheres is to the dark halos of galaxies. Many spiral galaxies have rotation curves for which v_{circ} is nearly constant over a large range in radius, as shown in Figure 6.2. The density of stars in the disk of a spiral galaxy generally falls exponentially with radius. At small radii, the disk of stars is self-gravitating, with little contribution from dark matter; we discuss self-gravitating disks in Section 6.5. At large radii, however, the orbiting stars of the disk can be thought of as test particles moving in a halo whose density ρ is provided mainly by dark matter.

The solar system is at a distance $r_\odot = 8.2\,\text{kpc}$ from the center of the Milky Way Galaxy. The circular velocity for stars in the solar neighborhood is $v_{\text{circ}} = 235\,\text{km s}^{-1}$. Normalized to these values, the density profile of a singular isothermal sphere can be written as

$$\rho(r) = \frac{v_{\text{circ}}^2}{4\pi G} \frac{1}{r^2} = 0.0152\,\text{M}_\odot\,\text{pc}^{-3} \left(\frac{v_{\text{circ}}}{235\,\text{km s}^{-1}}\right)^2 \left(\frac{r}{8.2\,\text{kpc}}\right)^{-2}. \tag{6.97}$$

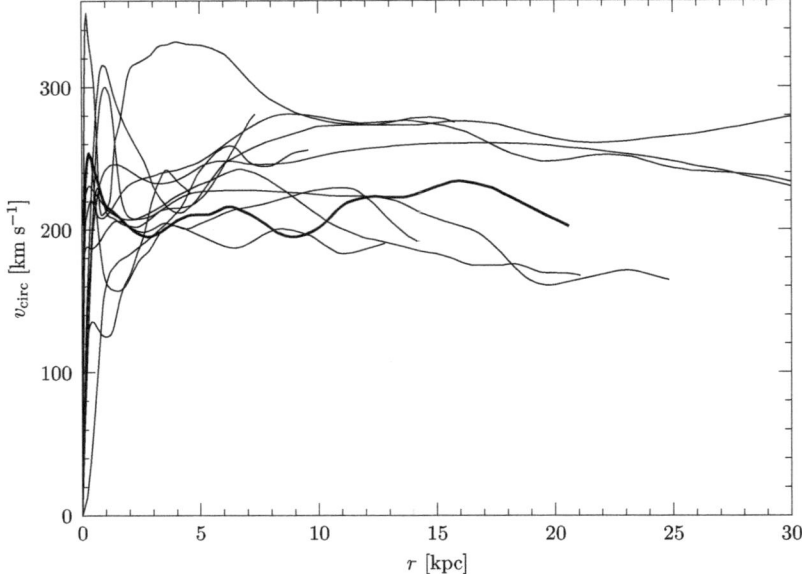

Figure 6.2 Rotation curves for spiral galaxies of class Sb; the heavy black line is the rotation curve of the Milky Way Galaxy, normalized to $r_\odot = 8\,\mathrm{kpc}$ and $v_{\mathrm{circ}} = 200\,\mathrm{km\,s^{-1}}$. [Sofue *et al.* 1999]

The enclosed mass is then

$$M(r) = \frac{v_{\mathrm{circ}}^2}{G} r = 1.05 \times 10^{11}\,\mathrm{M_\odot} \left(\frac{v_{\mathrm{circ}}}{235\,\mathrm{km\,s^{-1}}}\right)^2 \left(\frac{r}{8.2\,\mathrm{kpc}}\right). \qquad (6.98)$$

The total mass of the Milky Way depends on how far its roughly spherical, approximately isothermal halo of dark matter extends. If it has a maximum radius $r_{\mathrm{t}} \approx 4r_\odot \approx 33\,\mathrm{kpc}$, then its total mass $M_{\mathrm{t}} \sim 4 \times 10^{11}\,\mathrm{M_\odot}$. If the halo extends to $r_{\mathrm{t}} \approx 40r_\odot \approx 330\,\mathrm{kpc}$ (nearly halfway to the Andromeda Galaxy), its mass is a plump $M_{\mathrm{t}} \sim 4 \times 10^{12}\,\mathrm{M_\odot}$.

For radii less than the maximum radius r_{t}, the potential of a singular isothermal sphere is logarithmic,

$$\Phi(r) = -v_{\mathrm{circ}}^2 \left[1 - \ln(r/r_{\mathrm{t}})\right], \qquad (6.99)$$

resulting in an escape speed

$$v_{\mathrm{esc}}(r) = \sqrt{2} v_{\mathrm{circ}} \left[1 - \ln(r/r_{\mathrm{t}})\right]^{1/2}. \qquad (6.100)$$

If our galaxy's halo is large, with maximum radius $r_{\mathrm{t}} \approx 40r_\odot$, then the escape speed from the solar neighborhood is $v_{\mathrm{esc}} \approx 3.1 v_{\mathrm{circ}} \approx 720\,\mathrm{km\,s^{-1}}$. A skimpier halo, with $r_{\mathrm{t}} \approx 4r_\odot$, would lead to a local escape speed $v_{\mathrm{esc}} \approx 2.2 v_{\mathrm{circ}} \approx 510\,\mathrm{km\,s^{-1}}$, which is still large compared to the typical speed of stars in the solar neighborhood.

6.4.2 Cored Power-Law Profiles

Some stellar systems are observed to have cores of nearly constant stellar density. If the mass density traces the stellar density, it is sometimes useful to use a **cored power-law** profile for the mass density, with

$$\rho(r) = \frac{\rho_c}{(1 + r^2/r_c^2)^{\beta/2}}. \tag{6.101}$$

The corresponding mass profile is

$$M(r) = 4\pi \rho_c r_c^3 \int_0^{r/r_c} \frac{s^2 \, ds}{(1 + s^2)^{\beta/2}}. \tag{6.102}$$

Within the core ($r \ll r_c$), a Taylor series expansion for the density yields an enclosed mass

$$M(r \ll r_c) \approx M_c s^3 \left[1 - \frac{3\beta}{10} s^2 \right], \tag{6.103}$$

where $s \equiv r/r_c$ and

$$M_c \equiv \frac{4\pi}{3} \rho_c r_c^3. \tag{6.104}$$

Inside the nearly uniform density core, the circular speed represents nearly solid-body rotation, with

$$v_{\text{circ}}(r \ll r_c) \approx \left(\frac{GM_c}{r_c} \right)^{1/2} s \left[1 - \frac{3\beta}{20} s^2 \right]. \tag{6.105}$$

For $\beta > 3$, the system has a finite mass, without the need to impose a truncation radius r_t. Equation 6.102 yields a total mass for the system of

$$\frac{M_t}{M_c} = \frac{3\sqrt{\pi}}{4} \frac{\Gamma(\frac{\beta-3}{2})}{\Gamma(\frac{\beta}{2})}. \tag{6.106}$$

In 1911, Henry Plummer found that $\beta = 5$ gave a reasonable fit to the distribution of stars within a globular cluster. The resulting **Plummer profile**,

$$\rho_P(s) = \frac{\rho_c}{(1 + s^2)^{5/2}}, \tag{6.107}$$

has analytically tractable properties. Its enclosed mass is

$$M(s) = M_c \frac{s^3}{(1 + s^2)^{3/2}}, \tag{6.108}$$

leading to $M_t = M_c$ exactly. The gravitational potential is

$$\Phi_P(s) = -\frac{GM_c}{r_c} \frac{1}{(1 + s^2)^{1/2}}. \tag{6.109}$$

Because of the steep $\rho \propto r^{-5}$ decrease of the Plummer profile at large radii, its potential at $r \gg r_c$ is nearly identical to that of a point mass with $m = M_c$; this implies

$$v_{\rm esc}(r \gg r_c) \approx \sqrt{2}v_{\rm circ} \approx \left(\frac{2GM_c}{r}\right)^{1/2}. \tag{6.110}$$

The projected surface density of the Plummer profile is

$$\Sigma_{\rm P}(r) = \frac{M_c}{\pi r_c^2} \frac{1}{(1 + s^2)^2}. \tag{6.111}$$

It was by examining the surface brightness of nearby globular clusters such as ω Centauri that Plummer concluded that the Plummer profile gave an adequate fit to the number density of stars in globular clusters.[6]

6.4.3 Double Power-Law Profiles

Not every stellar system has a resolvable constant density core. In addition, not every cuspy stellar system is well fitted by a pure power law for its mass distribution. If a system's density makes a smooth transition from a shallow power law near its center to a steeper power law in its outer regions, a useful model is the **double power-law** profile,

$$\rho(r) = \frac{\rho_s}{(r/r_s)^\gamma (1 + r/r_s)^{\beta-\gamma}}. \tag{6.112}$$

When $r \gg r_s$, the density profile is $\rho \propto r^{-\beta}$; thus the total mass is finite when $\beta > 3$. Conversely, when $r \ll r_s$, the density is $\rho \propto r^{-\gamma}$, requiring $\gamma < 3$ to prevent a divergence in mass at the origin. When $\beta > 3 > \gamma$, the total mass of the double power-law model is finite, with

$$M_t = 4\pi\rho_s r_s^3 \frac{\Gamma(3 - \gamma)\Gamma(\beta - 3)}{\Gamma(\beta - \gamma)}. \tag{6.113}$$

The scaling radius r_s in Equation 6.112 is the radius at which the logarithmic slope in density is the average of the inner and outer slope, with

$$\left.\frac{d\ln\rho}{d\ln r}\right|_{r=r_s} = -\frac{1}{2}(\gamma + \beta). \tag{6.114}$$

Models with $\beta = 4$ and $\gamma < 3$, known as **Dehnen models**, are used to study the stellar distribution of elliptical galaxies. For a Dehnen model, the total mass is

$$M_t = \frac{4\pi}{3 - \gamma}\rho_s r_s^3, \tag{6.115}$$

---

[6] For ω Cen, Plummer found $r_c = 7.14$ arcmin; assuming a distance $d = 5.2$ kpc for ω Cen, this corresponds to $r_c = 11$ pc.

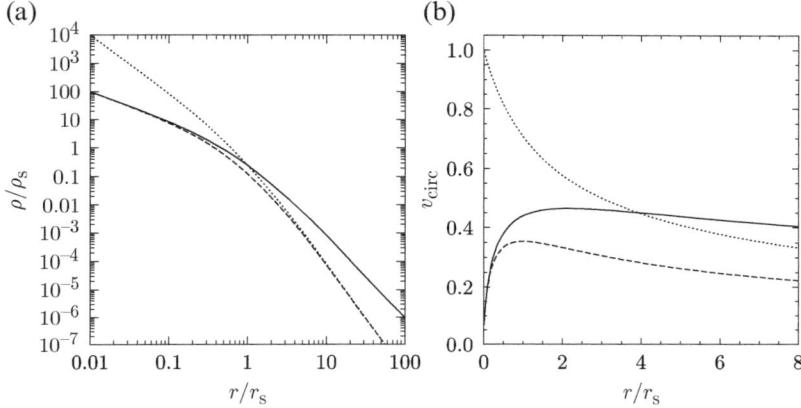

Figure 6.3 (a) Density profiles of a Jaffe model ($\beta = 4, \gamma = 2$: dotted line), a Hernquist model ($\beta = 4, \gamma = 1$: dashed line), and an NFW model ($\beta = 3, \gamma = 1$: solid line). (b) Circular speed within a Jaffe model (dotted), a Hernquist model (dashed), and an NFW model (solid). The circular speeds are in units of $(4\pi G\rho_s)^{1/2}r_s$.

and the enclosed mass is simply

$$M(s) = M_t \frac{s^{3-\gamma}}{(1+s)^{3-\gamma}}, \tag{6.116}$$

where $s \equiv r/r_s$. For a shallow central cusp ($0 < \gamma < 2$), a Dehnen model has potential

$$\Phi_D(s) = -\frac{GM_t}{(2-\gamma)a}\left[1 - \left(\frac{s}{1+s}\right)^{2-\gamma}\right] \qquad [0 < \gamma < 2]. \tag{6.117}$$

A Dehnen model with a $\gamma = 2$ core is also known as a Jaffe model; its potential (like that of a singular isothermal sphere) diverges logarithmically at the origin, with

$$\Phi_D(s) = -\frac{GM_t}{r_s}\ln\frac{1+s}{s} \qquad [\gamma = 2]. \tag{6.118}$$

The surface brightness of elliptical galaxies can be fitted fairly well by the projection of a Dehnen model with $\gamma = 2$ (a Jaffe model) or $\gamma = 1$ (called a Hernquist model).

In Figure 6.3(a), the normalized density profile ρ/ρ_s for a Jaffe model is shown as the dotted line; by comparison, a Hernquist model, with its shallower central cusp, is shown as the dashed line. Figure 6.3(b) shows the resulting circular velocity for the Jaffe profile (dotted line) and the Hernquist profile (dashed line). The circular speed is given in units of $(4\pi G\rho_s)^{1/2}r_s$.

Another application of the general double power-law model is to halos of dark matter. Navarro, Frenk, and White (1997) found that dark matter halos in numerical simulations were well fitted by a double power-law with $\gamma = 1$ for the inner slope and $\beta = 3$ for the outer slope. The resulting mass density profile, called an NFW model, is given by

$$\rho(r) = \frac{\rho_s}{(r/r_s)(1 + r/r_s)^2},\tag{6.119}$$

and is shown as the solid line in Figure 6.3(a). The enclosed mass for this profile can be written as

$$M(s) = 4\pi\rho_s r_s^3 A(s),\tag{6.120}$$

where the dimensionless radius is $s \equiv r/r_s$, and the dimensionless mass function is

$$A(s) \equiv \ln(1 + s) - \frac{s}{1 + s}.\tag{6.121}$$

In the limit of small and large radius,

$$A(s) = \begin{cases} \frac{1}{2}s^2(1 - \frac{4}{3}s) & [s \ll 1], \\ \ln s - 1 + 2/s & [s \gg 1]. \end{cases}\tag{6.122}$$

The resulting rotation curve for an NFW model is shown as the solid line in Figure 6.3(b). The circular speed in an NFW model has a maximum value of $v_{circ} \approx 1.64(G\rho_s)^{1/2}r_s$ at a radius $s = r/r_s \approx 2.163$. However, the circular speed is within 90% of the maximum value over the range $0.76 < s < 6.66$, producing a fairly flat rotation curve in this range.

The mass of an NFW model diverges logarithmically in the limit $r \to \infty$. However, a halo of dark matter has a natural truncation radius at its **virial radius** r_{vir}. A dark halo is a region of the universe that was initially slightly overdense, but which has collapsed and undergone relaxation to form a dense, gravitationally bound system that obeys the virial theorem (Section 6.1). However, outside a finite boundary, matter is still falling inward to the dark halo, and is not yet "virialized." In the spherical approximation, this boundary is a sphere whose radius equals r_{vir}. In practice, the virial radius is set equal to the radius r_{200} inside which the mean density of the dark halo is equal to 200 times the critical density ρ_c of the universe. Thus,

$$\bar{\rho}(r < r_{200}) = 200\rho_c(t) = 200\frac{3H(t)^2}{8\pi G},\tag{6.123}$$

where $H(t)$ is the Hubble parameter for the expansion of the universe. At the present day ($t = t_0$, $H(t) = H_0$) this requires

$$\bar{\rho}(r < r_{200}) = 200\frac{3H_0^2}{8\pi G} = 2.72 \times 10^{13}\, M_\odot\, \text{Mpc}^{-3}h_{70}^2,\tag{6.124}$$

where $h_{70} = H_0/70\,\text{km s}^{-1}\,\text{Mpc}^{-1}$.

If we truncate the NFW model at a radius r_{200}, its total mass is

$$M_t = 4\pi G\rho_s r_s^3 A(C),$$ (6.125)

where $C \equiv r_{200}/r_s$ is the **concentration parameter** of the halo. Numerical simulations of structure formation lead us to expect concentration parameters in the range $3 < C < 30$ for the dark halos of galaxies today, corresponding to $0.64 < A(C) < 2.5$. In some circumstances, it is useful to parameterize the NFW model in terms of the scaling density and radius (ρ_s and r_s) and the dimensionless radius $s \equiv r/r_s$. However, it can also be parameterized in terms of its total mass M_t, its concentration parameter C, and an alternative dimensionless radius $w \equiv r/r_{200}$ that runs from $w = 0$ to $w = 1$. For instance, we can write the NFW mass profile as

$$\frac{M(w)}{M_t} = \frac{A(Cw)}{A(C)},$$ (6.126)

and the equivalent circular velocity curve as

$$\frac{v_{\text{circ}}(w)}{v_{\text{circ}}(1)} = \left[\frac{A(Cw)}{wA(C)}\right]^{1/2}.$$ (6.127)

Using the definition of r_{200}, we can write the NFW density profile as

$$\frac{\rho(w)}{200\rho_{c,0}} = \frac{C^2}{3A(C)} \frac{1}{w(1 + Cw)^2}.$$ (6.128)

Figure 6.4(a) shows the density profile for NFW models with different concentration parameters, while Figure 6.4(b) shows the resulting rotation curves. Note that for the least concentrated halo shown, the virial radius at $r_{200} = 3r_s$ is not very far beyond the radius of maximum v_{circ}, at $r = 2.163r_s$.

6.5 Axisymmetric Systems

An infinitesimally thin, circular disk of stars is a rough approximation to the disks of spiral galaxies. If a thin disk is axisymmetric, then cylindrical coordinates are useful, with R being the distance from the center within the disk plane, and z being the distance perpendicular to the disk plane. Suppose that the disk has surface mass density $\Sigma(R)$. If the disk is infinitesimally thin, then Poisson's equation can be written as

$$\frac{1}{R}\frac{\partial}{\partial R}\left(R\frac{\partial\Phi}{\partial R}\right) + \frac{\partial^2\Phi}{\partial z^2} = 0 \qquad [z \neq 0].$$ (6.129)

If we try a separable solution for the potential, of the form

$$\Phi(R, z) = J(R)E(z),$$ (6.130)

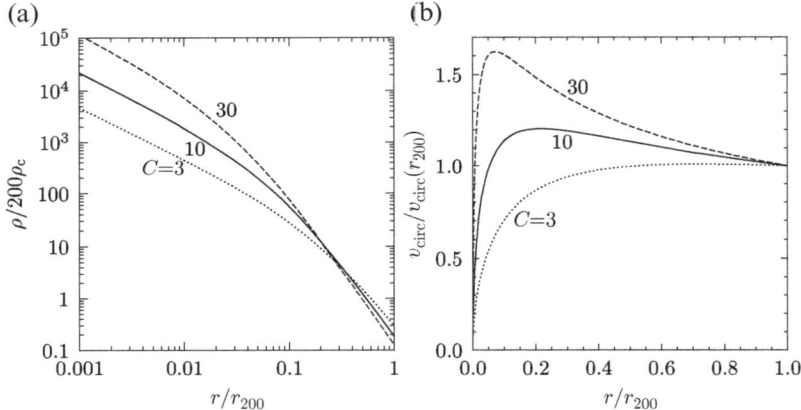

Figure 6.4 (a) Density profiles of an NFW model with concentration parameter $C =$ 3 (dotted line), $C = 10$ (solid), and $C = 30$ (dashed). The density is normalized to the mean density $\bar{\rho} = 200\rho_c$. (b) Circular speed within an NFW model with $C = 3$ (dotted line), $C = 10$ (solid), and $C = 30$ (dashed). The circular speed is normalized to its value at $r = r_{200}$.

then Poisson's equation (Equation 6.129) becomes

$$\frac{1}{JR}\frac{\partial}{\partial R}\left(R\frac{\partial J}{\partial R}\right) = -\frac{1}{E}\frac{\partial^2 E}{\partial z^2}. \tag{6.131}$$

The left side of Equation 6.131 is a function only of R; the right side is a function only of z. For the two sides to be equal for all values of R and z, they must equal a constant, which we can write as

$$\frac{1}{JR}\frac{\partial}{\partial R}\left(R\frac{\partial J}{\partial R}\right) = -\frac{1}{E}\frac{\partial^2 E}{\partial z^2} = -k^2. \tag{6.132}$$

The solution of Equation 6.132 for $E(z)$ is

$$E(z) \propto \exp(\pm kz). \tag{6.133}$$

If you are a fan of Bessel functions, you will have spotted that the solution for $J(R)$ is

$$J(R) \propto J_0(kR), \tag{6.134}$$

where J_0 is the cylindrical Bessel function of order zero. In the limits of small and large argument,

$$J_0(x) \approx \begin{cases} 1 - x^2/4 & [x \ll 1], \\ (2/\pi)^{1/2} x^{-1/2} \cos(x - \pi/4) & [x \gg 1]. \end{cases} \tag{6.135}$$

Figure 6.5 A small cylinder, whose end caps have area dA, sticks out above and below an infinitesimally thin disk.

For a disk of finite size, we want a solution for $\Phi(R, z)$ that goes to zero as $R \rightarrow \infty$ and $z \rightarrow \infty$. This picks out the solution

$$\Phi_k(R, z) = -2v_{k,0}^2 \exp(-k|z|)J_0(kR), \tag{6.136}$$

where the wavenumber k is real and positive; the depth of the potential is expressed here in terms of the escape speed $v_{k,0}$ from the disk's center.

Although the function Φ_k is a mathematically correct solution to Poisson's equation for a thin disk (Equation 6.129), it lacks one vital bit of information: the surface density $\Sigma(R)$. We can find the particular surface density Σ_k that gives rise to the potential Φ_k by using the divergence theorem. We choose a small area dA of the disk, and build a cylinder with end area dA that extends slightly above and below the disk, as shown in Figure 6.5. The divergence theorem states that for any vector function $\vec{F}$ within a region of volume V and surface area S,

$$\int (\vec{\nabla} \cdot \vec{F}) \, dV = \int (\vec{F} \cdot \hat{n}) \, dS, \tag{6.137}$$

where $\hat{n}$ is the unit vector normal to the surface. If we take the function $\vec{F}$ to be $\vec{\nabla}\Phi_k$ and the volume V to be that of our small cylinder, then Equation 6.137 becomes

$$\int (\nabla^2 \Phi_k) \, dV = \int_{\text{upper}} (\nabla\Phi_k \cdot \hat{n}) \, dA + \int_{\text{lower}} (\nabla\Phi_k \cdot \hat{n}) \, dA. \tag{6.138}$$

This equation includes terms for both the upper and lower end caps of the cylinder, but the height of the cylinder (and thus the area of its side) is assumed to be negligibly small. The left side of Equation 6.138, when we use Poisson's equation, is

$$\int (\nabla^2 \Phi_k) \, dV = 4\pi G \Sigma_k \, dA, \tag{6.139}$$

where Σ_k is the mass surface area of the disk at the location of dA. The right side of Equation 6.138, when we use the potential Φ_k (Equation 6.136), becomes

$$\int_{\text{upper}} (\nabla\Phi_k \cdot \hat{n}) \, dA + \int_{\text{lower}} (\nabla\Phi_k \cdot \hat{n}) \, dA = 2[2v_{k,0}^2 kJ_0(kR)] \, dA. \tag{6.140}$$

Substituting from Equations 6.139 and 6.140 into Equation 6.138, we find that

$$\Sigma_k(R) = \frac{v_{k,0}^2}{\pi G} k J_0(kR) \tag{6.141}$$

is the surface density distribution that gives rise to the potential Φ_k. At first glance, this result seems monumentally useless. The Bessel function $J_0(x)$ oscillates between positive and negative (Equation 6.135); however, physical reality dictates that the mass surface density of a disk must be non-negative. Thus, the function Σ_k cannot represent the surface density of an actual disk.

However, since $\Sigma_k \propto k J_0(kR)$ is a valid solution of Poisson's equation for any real, positive wavenumber k, it follows that any weighted integral of $k J_0(kR)$ over wavenumber will also provide a valid solution of Poisson's equation.[7] Thus, for any axisymmetric mass surface density $\Sigma(R)$, there exists a weighting function $S(k)$ such that

$$\Sigma(R) = \int_0^\infty S(k) J_0(kR) k \, dk. \tag{6.142}$$

We are now left with the task of determining the value of the function $S(k)$ for a known surface density profile $\Sigma(R)$. This becomes easier when we note from Equation 6.142 that $S(k)$ is the **Hankel transform** of $\Sigma(R)$. In case Hankel transforms aren't part of your everyday mathematical toolbox, the main thing you need to know is that Hankel transforms are to cylindrical Bessel functions as Fourier transforms are to sinusoids. That is, if

$$f(R) = \int_0^\infty g(k) J_\nu(kR) k \, dk \tag{6.143}$$

then it follows from the properties of Bessel functions that

$$g(k) = \int_0^\infty f(R) J_\nu(kR) R \, dR, \tag{6.144}$$

and $f(R)$ and $g(k)$ are Hankel transforms (of order ν) of each other. Thus, from Equation 6.142, we can deduce that

$$S(k) = \int_0^\infty \Sigma(R) J_0(kR) R \, dR. \tag{6.145}$$

Notice the normalization we have chosen: if $\Sigma(R)$ has units of $M_\odot \, pc^{-2}$, then $S(k)$ has units of $M_\odot$.

Once you have determined $S(k)$ for the surface density $\Sigma(R)$ in question, other properties of the disk can be computed. For instance, the potential $\Phi(R, z)$ of the surface density $\Sigma(R)$ is given by

$$\Phi(R, z) = -2\pi G \int_0^\infty S(k) \exp(-k|z|) J_0(kR) \, dk. \tag{6.146}$$

---

[7] This follows from the linear nature of Poisson's equation.

The circular speed of stars in the disk is then

$$v_{\text{circ}}^2 = R\frac{\partial \Phi}{\partial R}\bigg|_{z=0} = -2\pi GR \int_0^\infty S(k)\frac{dJ_0(kR)}{dR}\,dk. \tag{6.147}$$

Since Bessel functions have the general property that

$$\frac{d}{dx}[J_\nu(x)x^\nu] = -\frac{J_{\nu+1}(x)}{x^\nu}, \tag{6.148}$$

Equation 6.147 can be written more compactly as

$$v_{\text{circ}}^2 = 2\pi GR \int_0^\infty S(k)J_1(kR)k\,dk, \tag{6.149}$$

where J_1 is the cylindrical Bessel function of order one. For a thin disk, it is not generally true that $v_{\text{circ}}^2 = GM(R)/R$. Since we found $S(k)$ by integrating over all values of R, the circular speed at a distance R_0 from the disk center can depend on the mass distribution at $R > R_0$.

As an example, take a thin disk with surface mass density

$$\Sigma(R) = \Sigma_t\left(\frac{R_t}{R}\right) \tag{6.150}$$

out to a truncation radius R_t. Such a disk is known as a **Mestel disk**, after Leon Mestel, who discussed its properties in 1963. The enclosed mass for a Mestel disk is

$$M(R) = M_t\left(\frac{R}{R_t}\right), \tag{6.151}$$

where $M_t = 2\pi\Sigma_t R_t^2$. It is convenient to describe the properties of the Mestel disk in terms of a parameter

$$v_M^2 \equiv \frac{GM(R)}{R} = \frac{GM_t}{R_t}. \tag{6.152}$$

This "Mestel speed" is what the circular speed would be in a Mestel disk if the mass were spherically distributed. In terms of the Mestel speed, the surface density and enclosed mass of the Mestel disk are (for $R \le R_t$)

$$\Sigma(R) = \frac{v_M^2}{2\pi G}\frac{1}{R}, \qquad M(R) = \frac{v_M^2}{G}R. \tag{6.153}$$

The function $S(k)$ for a Mestel disk is, from Equation 6.145,

$$S(k) = \frac{v_M^2}{2\pi G}\int_0^{R_t} J_0(kR)\,dR = \frac{v_M^2}{2\pi G}\frac{1}{k}\int_0^{kR_t} J_0(x)\,dx. \tag{6.154}$$

The cylindrical Bessel functions are normalized so that

$$\int_0^\infty J_\nu(x)\,dx = 1. \tag{6.155}$$

This means that in the limit of an *infinite* Mestel disk, which has $R_t \to \infty$, the function $S(k)$ is

$$S_{IM}(k) = \frac{v_M^2}{2\pi G} \frac{1}{k}. \tag{6.156}$$

For a finite Mestel disk, the function is

$$S_{FM}(k) \approx \frac{v_M^2}{2\pi G} \frac{1}{k} \left[1 + O\left([kR_t]^{-1/2}\right)\right] \tag{6.157}$$

when $k \gg R_t^{-1}$. For an infinite Mestel disk, the circular speed, found from Equation 6.149, turns out to be delightfully simple:

$$v_{circ}^2 = \frac{v_M^2}{2\pi G} \left[2\pi G R \int_C^\infty J_1(kR)\, dk\right] = v_M^2. \tag{6.158}$$

Thus, a self-gravitating infinite Mestel disk is the unusual case of a thin disk that *does* have $v_{circ}^2 = GM(R)/R$. A finite Mestel disk has a circular speed

$$v_{circ}^2 = v_M^2 \left[1 + O\left([R/R_t]^{1/2}\right)\right] \tag{6.159}$$

when $R \ll R_t$.

Another interesting case is that of an exponential disk, with surface mass density

$$\Sigma(R) = \Sigma_0 \exp(-R/R_d). \tag{6.160}$$

The enclosed mass for such a disk is finite, with

$$M(R) = M_d \left[1 - e^{-R/R_d}(1 + R/R_d)\right], \tag{6.161}$$

where the total disk mass is $M_d = 2\pi \Sigma_0 R_d^2$. Figure 6.6 shows the circular speed v_{circ} for such an exponential thin disk. The solid curve shows v_{circ} calculated correctly, with the use of Hankel transforms; the circular speed peaks at $R \approx 2.1 R_d$, where $v_{circ}^2 = 0.387 GM_d/R_d$. The dashed line in Figure 6.6 shows the "spherical disk" approximation, which assumes (incorrectly for a disk other than the infinite Mestel disk) that $v_{circ}^2 = GM(R)/R$. In this approximation, the peak in the circular speed is shifted inward, to $R \approx 1.8 R_d$, where the spherical disk approximation yields $v_{circ}^2 = 0.298 GM_d/R_d$. In yet another leap of approximation, the dotted line in Figure 6.6 shows the "point disk" approximation, in which all the mass M_d of the disk is placed at the origin. At $R < R_d$, this is a truly wretched approximation. However, at $R > 2R_d$, the Keplerian point disk approximation actually gives a closer estimate of $v_{circ}(R)$ than the spherical disk approximation does.[8]

Sometimes the approximation of an infinitesimally thin disk is useful. However, we can obtain additional information about real stellar disks (like the

---

[8] At $R = 2.66 R_d$ and in the limit $R \gg R_d$, the Keplerian approximation yields the correct circular speed; but then, even a broken clock is right twice a day.

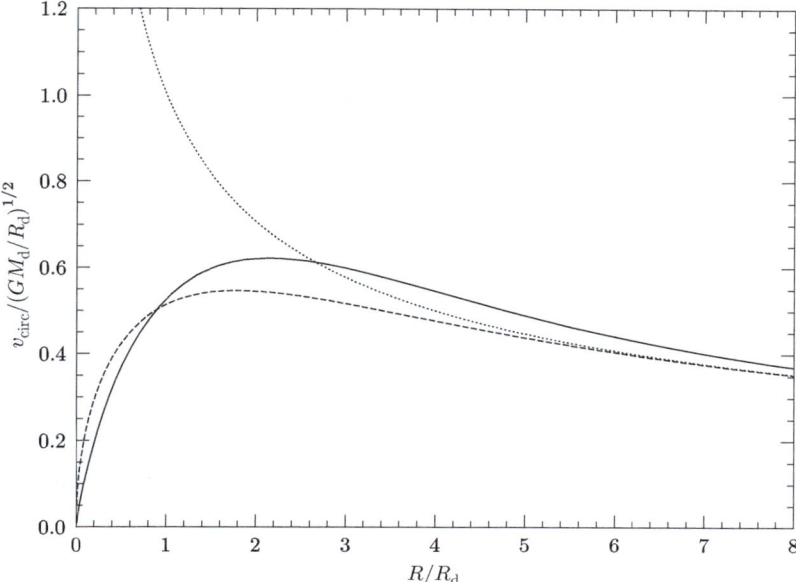

Figure 6.6 The true circular speed of an exponential disk (solid line), compared with the "spherical disk" approximation (dashed) and the "point disk" approximation (dotted).

disk of our own galaxy) by acknowledging that disks have a small but finite thickness. In this context, a "thin disk" is not one with a scale height $h_z = 0$ in the z direction; instead, it is one with $h_z \ll R_d$, where R_d is the disk scale length in the R direction. In general, an axisymmetric disk must obey a Jeans equation (Equation 6.69),

$$\frac{1}{R}\frac{\partial}{\partial R}\left(Rn_\star\sigma_{Rz}^2\right) + \frac{\partial}{\partial z}\left(n_\star\sigma_{zz}^2\right) = -n_\star\frac{\partial\Phi}{\partial z}. \tag{6.162}$$

For a thin disk, however, gradients in the R direction are smaller than those in the z direction. Thus, we may make the approximation

$$\frac{1}{n_\star}\frac{\partial}{\partial z}\left(n_\star\sigma_{zz}^2\right) = -\frac{\partial\Phi}{\partial z} \tag{6.163}$$

for a disk that is thin (but not infinitesimally thin). Since Equation 6.163 is based on a continuity equation, the number density $n_\star$ can be the number density of anything that is conserved. It could be the number density of stars, or white dwarfs, or self-gravitating spherical cows (assuming spherical cow conservation).

In general, Poisson's equation for an axisymmetric system is

$$\frac{1}{R}\frac{\partial}{\partial R}\left(R\frac{\partial\Phi}{\partial R}\right)+\frac{\partial^2\Phi}{\partial z^2}=4\pi G\rho(R,z). \tag{6.164}$$

Near the midplane of a thin disk, we can write this as

$$\frac{1}{R}\frac{\partial v_{\text{circ}}}{\partial R}+\frac{\partial^2\Phi}{\partial z^2}\approx 4\pi G\rho(R,z). \tag{6.165}$$

When the rotation curve of the disk is flat (with v_{circ} approximately constant), Poisson's equation finally reduces to

$$\frac{\partial^2\Phi}{\partial z^2}\approx 4\pi G\rho(R,z). \tag{6.166}$$

Combining Equations 6.163 and 6.166, we find an equation that links the total mass density ρ with the number density $n_\star$ and velocity dispersion σ_{zz} of our population of tracer stars:

$$\rho=-\frac{1}{4\pi G}\frac{\partial}{\partial z}\left[\frac{1}{n_\star}\frac{\partial}{\partial z}(n_\star\sigma_{zz}^2)\right]. \tag{6.167}$$

The good news is that we can find the mass density ρ, including all the dark matter in the disk, if we know $n_\star$ and σ_{zz} as a function of z. The bad news is that we need to know $n_\star$ and σ_{zz} well enough to differentiate twice.

As an example, consider the thin disk component of our own galaxy. At the Sun's location, 8.2 kpc from the galactic center, the thin disk has a scale length $R_d\sim 3$ kpc in the R direction. By contrast, the distribution of thin-disk stars in the z direction can be modeled as

$$n_\star(z)=n_0\,\text{sech}^2\left(\frac{z}{2h_z}\right), \tag{6.168}$$

with $h_z\sim 0.2$ kpc. In the limit $|z|\gg h_z$, Equation 6.168 yields an exponentially falling density, with $n_\star\propto\exp(-|z|/h_z)$; however, in the limit $|z|\ll h_z$, $n_\star\propto 1-0.25(z/h_z)^2$, avoiding a discontinuity in $dn_\star/dz$ at the midplane of the disk. If σ_{zz} is assumed to be constant with z, then Equation 6.167 tells us that the mass density at midplane is

$$\rho(0)=\frac{\sigma_{zz}^2}{8\pi Gh_z^2}\approx 0.093\,\text{M}_\odot\,\text{pc}^{-3}\left(\frac{\sigma_{zz}^2}{20\,\text{km s}^{-1}}\right)^2\left(\frac{h_z}{0.2\,\text{kpc}}\right)^{-2}, \tag{6.169}$$

scaling to a plausible scale height and velocity dispersion for stellar populations in the thin disk of our galaxy.

The use of the Jeans equations (or the equivalent) to find the mass density near the Sun's location is a long-standing problem in astronomy. In a classic 1932 paper, Jan Oort used all the stellar information available at the time, coming to the conclusion "the most probable value of the total density near the Sun is thus .092 suns/pc^3, or 6.3×10^{-24} g/cm^3. The probable error of this quantity is estimated to be about 20%." In the decades that followed, although astronomers had access to more data, with improved stellar distances, they realized that the assumption of an axisymmetric, steady-state thin disk limits the usefulness of approximations such as that of Equation 6.167. However, even a $\pm 20\%$ approximation leads to the conclusion that the amount of dark matter locally, in the midplane of our galaxy's disk, must be small compared to the amount of visible matter.

Exercises

6.1 Newton's first shell theorem implies that a Dyson sphere (a rigid spherical shell of constant surface density centered on a star) is gravitationally stable. However, in 1856, James Clerk Maxwell famously showed that a rigid ring centered on a planet is unstable. Let's follow Maxwell's reasoning.

(a) A thin rigid ring of radius r is centered on a planet of mass m_1. The mass per unit length of the ring is constant, producing a total ring mass $m_2 \ll m_1$. Show mathematically that if the rigid ring is perfectly centered on the planet, it is in equilibrium, experiencing no net force.

(b) Show that if you displace the ring by a small distance δr in the plane defined by the ring, its displacement relative to the central planet grows with time.

(c) In the limit $\delta r \ll r$, what is the timescale τ for the growth of δr? [*Hint:* Linear perturbation theory is your friend.]

6.2 The stars in the thin disk of our galaxy have a velocity dispersion σ_{zz} that increases slightly with distance from the midplane. A reasonable fit is

$$\sigma_{zz} = \sigma_0 \left[1 + \frac{|z|}{h_\sigma} \right], \tag{6.170}$$

where $h_\sigma \sim 2\,\mathrm{kpc}$. Using this more realistic velocity dispersion, would your estimate of the midplane mass density $\rho(0)$ increase or decrease relative to the estimate of Equation 6.169, which assumes constant σ_{zz}? By what fractional amount would your estimate of $\rho(0)$ change, expressed as a function of h_z/h_σ?

6.3 In 1933, Fritz Zwicky famously suggested that the mass of the Coma cluster of galaxies could be dominated by dark matter. Zwicky based his argument on the observed radial velocities of eight galaxies in the Coma cluster: these velocities were $v_r = 8500, 7900, 7600, 7000, 6900, 6700, 6600,$ and $5100\,\mathrm{km\,s^{-1}}$.

(a) What are the mean and standard deviation of these eight radial velocities? Assuming $H_0 = 70\,\mathrm{km\,s^{-1}\,Mpc^{-1}}$, what is the distance to the Coma cluster?

(b) The half-light radius of the Coma cluster, in angular units, is $r_{hl} = 0.85°$. Zwicky's best estimate of the distance to the Coma cluster was $d = 14\,\mathrm{Mpc}$. Compute the virial mass of the Coma cluster using Zwicky's distance.

(c) Using the distance to the Coma cluster that you computed in part (a), recompute the virial mass of the Coma cluster.

(d) Zwicky acknowledged that the lowest-velocity galaxy in the sample of eight (with $v_r = 5100\,\mathrm{km\,s^{-1}}$) might be a foreground galaxy. Omitting this galaxy from the sample, what is your revised virial mass estimate for the Coma cluster?

6.4   Suppose that the stars in the disk of our galaxy are on coplanar circular orbits around the galactic center, with orbital angular speed $\Omega(r)$. From the Sun's location at a distance r_0 from the galactic center, we observe a number of nearby stars. For each star, we measure its distance d from the Sun, its radial velocity v_r relative to the Sun, its proper motion μ'', and its galactic longitude ℓ. (From the Sun's location, galactic longitude is defined so that $\ell = 0$ in the direction of the galactic center, and $\ell = \pi/2 = 90°$ in the direction of the Sun's orbital motion.)

(a) Show that for stars with $d \ll r_0$, the radial velocity obeys the relation $v_r = Ad\sin(2\ell)$, where

$$A = -\frac{1}{2}\left[r\frac{d\Omega}{dr}\right]_{r=r_0}. \qquad (6.171)$$

(b) Show that for stars with $d \ll r_0$, the proper motion obeys the relation $\mu'' = B + A\cos(2\ell)$, where

$$B = -\left[\Omega + \frac{1}{2}r\frac{d\Omega}{dr}\right]_{r=r_0}. \qquad (6.172)$$

(c) The constants A and B are known as the **Oort constants**. The best current values for the Oort constants are $A = 15.1\,\mathrm{km\,s^{-1}\,kpc^{-1}}$ and $B = -13.4\,\mathrm{km\,s^{-1}\,kpc^{-1}}$. What are the resulting values for Ω and $d\Omega/dr$ at the Sun's location? What is v_{circ} at the Sun's location, assuming $r_0 = 8.2\,\mathrm{kpc}$?

6.5   A spiral galaxy like our own consists of a small rotating disk of stars and gas within a larger, more slowly rotating halo of dark matter. Let's see how the relative sizes and rotation speeds are linked.

(a) Approximate the dark halo as a singular isothermal sphere with truncation radius r_t and total mass M_t. Assuming the halo is in virial equilibrium, what is its total energy E_t in terms of r_t and M_t?

(b) The net angular momentum J_t of a dark halo is often expressed in terms of a dimensionless spin parameter λ, defined as

$$\lambda \equiv J_t |E_t|^{1/2} G^{-1} M_t^{-5/2}. \tag{6.173}$$

Write an expression for J_t in terms of λ, r_t, and M_t for the dark halo.

(c) Suppose that the central disk of stars is a thin exponential disk with scale length R_d and mass M_d (Equation 6.161). If the stars are on circular orbits (with no "wrong-way" stars), what is the net angular momentum J_d of the disk, assuming that the dark halo dominates the motion of stars?

(d) In the early universe, dark halos were tidally torqued by neighboring halos; during this process, the baryonic gas destined to become stars was still well-mixed with the dark matter. Thus, if the disk-to-halo mass ratio is $M_d/M_t = f$, we also expect $J_d/J_t = f$. With this assumption, find the disk scale length R_d in terms of the halo properties (such as M_t, r_t, and λ).

(e) The disk of our galaxy has $M_d \approx 7 \times 10^{10}\,M_\odot$ and $R_d \approx 2.2\,\text{kpc}$. Assuming a dark halo with $v_{\text{circ}} = 235\,\text{km s}^{-1}$, and scaling to a truncation radius $r_t = 100\,\text{kpc}$, what is the value of λ? Simulations of halo formation predict $\lambda \approx 0.05$; would this value of λ argue for a large halo ($r_t > 100\,\text{kpc}$) or a small halo?

7

Orbits

When night comes
I stand on the steps and listen;
the stars cluster in the garden
and I stand, out in the darkness.

Edith Södergran (1892–1923)
"Stjärnorna [The Stars]" [1916] (tr. David Barrett)

Suppose that a test mass moves in some potential $\Phi(\vec{r}, t)$. For many problems in stellar dynamics, it is useful to know the orbit of the test mass. Is it bound, with a speed less than the local escape speed, or is it unbound? If it is bound, then is its orbit a closed curve, such as an ellipse, or is it a space-filling curve? In Section 6.5, we approximated the orbits of stars in a disk galaxy as being perfectly circular. However, no orbit is perfect; we will need a way to describe a star's deviations from a circular orbit.

In classical physics, as we have seen, it is often helpful to identify the conserved quantities of a system. For orbital motion, a **constant of motion** is a function $C(\vec{r}, \vec{v}, t)$ that is constant as a test mass moves along its orbit. For instance, if a point mass is on a circular orbit of radius a in a Keplerian potential $\Phi_K(r) = -\mu/r$, it moves at constant angular speed $\Omega = \mu^{1/2} a^{-3/2}$ (Equation 2.2). Thus, one constant of motion for this orbit is $\theta_0 = \theta(t) - \Omega t$, where $\theta(t)$ is the azimuthal angle of the test mass. More restrictively, an **integral of motion** is a function $I(\vec{r}, \vec{v})$ that has no explicit time dependence and is constant as a test mass moves along its orbit. For instance, if a point mass moves freely in a Keplerian potential, its specific energy $\epsilon = v^2/2 - \mu/r$ is an integral of motion; so is the specific angular momentum $\vec{j} = \vec{r} \times \vec{v}$.

7.1 Orbits in Spherical Potentials

Let's start with the simple case of a test particle moving in a static spherical potential $\Phi(r)$. The equation of motion is then

$$\frac{d^2\vec{r}}{dt^2} = -\frac{d\Phi}{dr}\hat{r}, \tag{7.1}$$

indicating that the particle is moving under the influence of a central force. As shown in Section 1.2, this means that the specific angular momentum of the test particle,

$$\vec{j} = \vec{r} \times \frac{d\vec{r}}{dt}, \tag{7.2}$$

is a conserved quantity. Thus, $\vec{j}$ is an integral of motion for all static spherical potentials, not merely Keplerian potentials.

Since $\vec{j}$ is an integral of motion, the orbit of the test mass is confined to an orbital plane. Within this plane, it is convenient to use polar coordinates (r, θ) with the origin at the potential's center. In these coordinates, the acceleration from a central force is purely radial, and can be expressed as (Equation 1.35)

$$\frac{d^2\vec{r}}{dt^2} = \left[\frac{d^2r}{dt^2} - \frac{j^2}{r^3}\right]\hat{r}, \tag{7.3}$$

where $j \equiv |\vec{j}|$ is the magnitude of the specific angular momentum. Combining Equations 7.1 and 7.3, the equation of motion in the radial direction is

$$\frac{d^2r}{dt^2} - \frac{j^2}{r^3} = -\frac{d\Phi}{dr}. \tag{7.4}$$

Thus, spherical potentials can support circular orbits. For a test mass to be on a circular orbit with radius $r = r_c$, its specific orbital angular momentum must be given by the relation

$$j^2 = \left[r^3\frac{d\Phi}{dr}\right]_{r=r_c} \tag{7.5}$$

However, orbits in an arbitrary spherical potential are not required to be circular.

Multiplying Equation 7.4 by dr/dt and integrating over time, we find an equation of energy conservation:

$$\frac{1}{2}\left(\frac{dr}{dt}\right)^2 + \frac{1}{2}\frac{j^2}{r^2} + \Phi(r) = \epsilon = \text{constant.} \tag{7.6}$$

In terms of the integrals of motion ϵ and j, the radial velocity of the particle is

$$\frac{dr}{dt} = \pm\sqrt{2\epsilon - 2\Phi(r) - j^2/r^2}. \tag{7.7}$$

On a bound orbit, with $\epsilon < 0$, the radial velocity equals zero at periapsis ($r = r_{pe}$) and at apoapsis ($r = r_{ap} > r_{pe}$).

The radial period of the test mass (that is, the time it takes to go from periapsis to apoapsis and back again) is

$$\mathcal{P}_r = 2 \int_{r_{pe}}^{r_{ap}} \left(\frac{dr}{dt}\right)^{-1} dr = 2 \int_{r_{pe}}^{r_{ap}} \frac{dr}{\sqrt{2\epsilon - 2\Phi(r) - j^2/r^2}}, \tag{7.8}$$

making use of Equation 7.7. The change in azimuthal angle θ during one radial period is

$$\Delta\theta = 2 \int_{r_{pe}}^{r_{ap}} \frac{d\theta}{dr} dr = 2 \int_{r_{pe}}^{r_{ap}} \frac{d\theta}{dt} \left(\frac{dr}{dt}\right)^{-1} dr. \tag{7.9}$$

Since $d\theta/dt = j/r^2$ for a central force, this can be written as

$$\Delta\theta = 2j \int_{r_{pe}}^{r_{ap}} \frac{dr}{r^2 \sqrt{2\epsilon - 2\Phi(r) - j^2/r^2}}. \tag{7.10}$$

The azimuthal period of the test mass (that is, the time it takes to go through 2π radians in θ) is

$$\mathcal{P}_\theta = \frac{2\pi}{\Delta\theta} \mathcal{P}_r. \tag{7.11}$$

In general, $2\pi/\Delta\theta$ is not a rational number; thus, $\mathcal{P}_\theta/\mathcal{P}_r$ will not be rational either. If this is the case, then bound orbits will not constitute closed curves.[1]

To determine $r(\theta)$ for an orbit in a spherical potential, start with the radial equation of motion (Equation 7.4), and make the substitution

$$\frac{d}{dt} = \frac{j}{r^2} \frac{d}{d\theta}, \tag{7.12}$$

which follows from angular momentum conservation. This changes the radial equation of motion to the form

$$\frac{j^2}{r^2} \frac{d}{d\theta} \left(\frac{1}{r^2} \frac{dr}{d\theta}\right) - \frac{j^2}{r^3} = -\frac{d\Phi}{dr}. \tag{7.13}$$

This equation takes a simpler appearance with the substitution $u \equiv 1/r$, which yields

$$\frac{d^2u}{d\theta^2} + u = -\frac{1}{j^2} \frac{d\Phi}{du}. \tag{7.14}$$

We've seen this equation before (as Equation 1.37) in the special case of a Keplerian potential, with $\Phi_K = -\mu/r = -\mu u$. Equation 7.14 has an analytic solution in the Keplerian case; however, for an arbitrary spherical potential $\Phi(r)$, it must be numerically integrated for an assumed angular momentum j.

There is one potential, in addition to the Keplerian potential, for which there are interesting analytic orbits. This is the **harmonic oscillator** potential.

---

[1] A Keplerian potential, $\Phi_K = -\mu/r$, is a special case for which $\mathcal{P}_\theta = \mathcal{P}_r$.

Suppose that a spherical stellar system has a central mass density ρ_c that is constant inside a radius r_c. The potential at radii $r < r_c$ then has the quadratic form

$$\Phi(r) = \frac{2\pi}{3} G\rho_c r^2 + \Phi_0, \tag{7.15}$$

where the value of the central potential Φ_0 depends on the distribution of mass outside the core.

If a test mass remains within the constant-density core, its equation of motion is (Equation 7.1)

$$\frac{d^2\vec{r}}{dt^2} = -\frac{d\Phi}{dr}\hat{r} = -\frac{4\pi}{3}G\rho_c\vec{r}. \tag{7.16}$$

For the harmonic oscillator potential, it is most illuminating to use a Cartesian coordinate system in the orbital plane, with $x = r\cos\theta$ and $y = r\sin\theta$. The equation of motion in these coordinates is then

$$\frac{d^2x}{dt^2} = -\omega_0^2 x, \qquad \frac{d^2y}{dt^2} = -\omega_0^2 y, \tag{7.17}$$

where the angular frequency of the harmonic oscillator is

$$\omega_0 = \left(\frac{4\pi}{3}G\rho_c\right)^{1/2} = 0.137\,\mathrm{Myr}^{-1}\left(\frac{\rho_c}{1\,\mathrm{M_\odot\,pc^{-3}}}\right)^{1/2}. \tag{7.18}$$

The solution to the equation of motion is

$$x(t) = x_0\cos(\omega_0 t), \tag{7.19}$$

$$y(t) = y_0\cos(\omega_0 t - \Delta). \tag{7.20}$$

We may assume, without loss of generality, that the amplitudes x_0 and y_0 are non-negative, and that the phase shift lies in the range $-\pi < \Delta \leq \pi$. We have also taken the mild liberty of setting $t = 0$, the arbitrary time origin, to be an instant when $x = x_0$. Since the frequency ω_0 is the same in the x and y directions, all orbits in the harmonic oscillator potential are *closed* orbits. In general, the orbit has a specific angular momentum $j = \omega_0 x_0 y_0 \sin\Delta$. In the special case $\Delta = 0$ or π, the harmonic oscillator orbit is thus purely radial, with apoapsis at $r_{\mathrm{ap}} = (x_0^2 + y_0^2)^{1/2}$, as shown in Figure 7.1. When $\Delta = \pm\pi/2$, the harmonic oscillator orbit is an ellipse whose primary axes align with the x and y axes; the semimajor axis is the larger of x_0 and y_0 while the semiminor axis is the smaller of x_0 and y_0. For arbitrary values of the phase shift Δ, the orbit is an ellipse whose primary axes are rotated through an angle θ_r relative to the x and y axes, where

$$\tan(2\theta_r) = \frac{2x_0 y_0 \cos\Delta}{x_0^2 - y_0^2}. \tag{7.21}$$

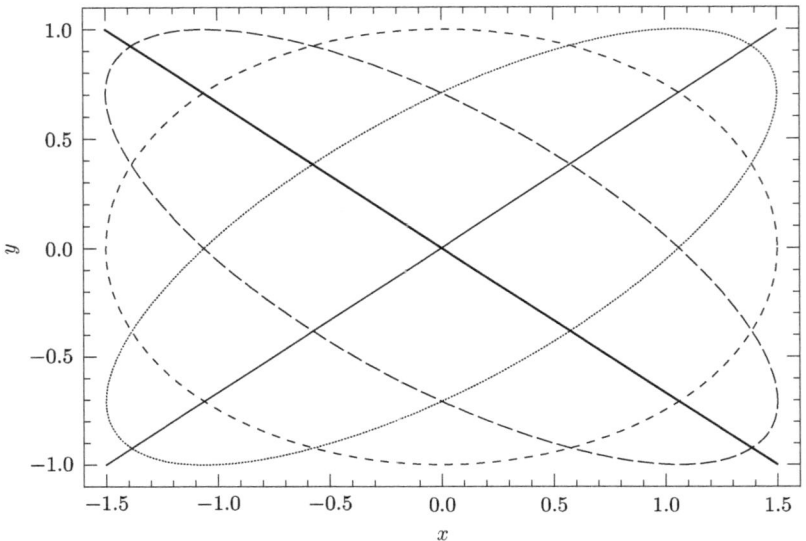

Figure 7.1 Orbits in a harmonic oscillator potential with $x_0 = 1.5$ and $y_0 = 1$. Light solid line: $\Delta = 0$. Dotted: $\Delta = \pi/4$. Short dashed: $\Delta = \pi/2$. Long dashed: $\Delta = 3\pi/4$. Heavy solid: $\Delta = \pi$.

Notice from Figure 7.1 that bound orbits in a harmonic oscillator potential are ellipses whose *centers* are located at the center of the potential. This is in contrast to bound orbits in a Keplerian potential, whose *foci* are located at the center of the potential.

In the harmonic oscillator potential, the test mass completes one orbit during one azimuthal period,

$$\mathcal{P}_\theta = \frac{2\pi}{\omega_0} = \sqrt{3\pi}(G\rho_c)^{1/2} = 45.8\,\mathrm{Myr}\left(\frac{\rho_c}{1\,M_\odot\,\mathrm{pc}^{-3}}\right)^{-1/2}. \qquad (7.22)$$

During this time, the test mass goes through periapsis twice and apoapsis twice. Thus, the radial period is

$$\mathcal{P}_r = \frac{1}{2}\mathcal{P}_\theta = \frac{\sqrt{3\pi}}{2}(G\rho_c)^{1/2} = 22.9\,\mathrm{Myr}\left(\frac{\rho_c}{1\,M_\odot\,\mathrm{pc}^{-3}}\right)^{-1/2}. \qquad (7.23)$$

We can compare the harmonic oscillator to the Keplerian potential, for which $\mathcal{P}_\theta = \mathcal{P}_r = 2\pi(a^3/\mu)^{1/2}$. For a test mass on a circular Keplerian orbit, the mean density within the orbit is

$$\bar{\rho}(r) = \frac{3}{4\pi}\frac{m}{r^3}, \qquad (7.24)$$

where m is the mass of the primary. Thus, the Keplerian periods can be written as

$$\mathcal{P}_\theta = \mathcal{P}_r = \sqrt{3\pi}(G\bar{\rho})^{-1/2}. \tag{7.25}$$

Speaking very generally, if gravity is the dominant force in a problem, the timescale on which things change will be proportional to the **dynamical time**

$$t_{\text{dyn}} \equiv (G\bar{\rho})^{-1/2} = 14.9\,\text{Myr} \left(\frac{\bar{\rho}}{1\,\text{M}_\odot\,\text{pc}^{-3}}\right)^{-1/2}. \tag{7.26}$$

If a non-rotating star is suddenly deprived of its internal pressure support, for instance, it will collapse to a black hole on a time equal to its freefall time,

$$t_{\text{ff}} = \left(\frac{3\pi}{32}\right)^{1/2} t_{\text{dyn}} = 29.5\,\text{min} \left(\frac{\bar{\rho}}{1.41\,\text{g cm}^{-3}}\right)^{-1/2}, \tag{7.27}$$

scaling to the Sun's mean density, $\rho_\odot = 1.41\,\text{g cm}^{-3}$. As another example, in an Einstein–de Sitter universe (expanding, flat, and matter-dominated), the time since the Big Bang is

$$t = \left(\frac{1}{6\pi}\right)^{1/2} t_{\text{dyn}} = 0.36\,\text{Gyr} \left(\frac{\bar{\rho}}{10^{14}\,\text{M}_\odot\,\text{Mpc}^{-3}}\right)^{-1/2}, \tag{7.28}$$

scaling to the mean density at the era of Cosmic Dawn,[2] when our universe was matter-dominated and the first galaxies were starting to pour out light.

7.2 Orbits in Axisymmetric Potentials

Spherical symmetry is an attractively simple approximation, but not always an adequate approximation. For example, in looking at the orbit of the Sun (and other stars in our galaxy's disk), it is important to take into account the flattening of the disk potential. Let's now look at static axisymmetric potentials. In cylindrical coordinates, such a potential can be written as $\Phi(R, z)$, where R is the distance from the axis of rotational symmetry, and z is the distance from a reference plane perpendicular to that axis. The equation of motion for an axisymmetric system is

$$\frac{d^2\vec{r}}{dt^2} = -\frac{\partial\Phi}{\partial R}\hat{R} - \frac{\partial\Phi}{\partial z}\hat{z}. \tag{7.29}$$

Expressed in cylindrical coordinates, the acceleration vector is

$$\frac{d^2\vec{r}}{dt^2} = \left[\frac{d^2R}{dt^2} - R\left(\frac{d\theta}{dt}\right)^2\right]\hat{R} + \left[\frac{1}{R}\frac{d}{dt}\left(R^2\frac{d\theta}{dt}\right)\right]\hat{\theta} + \frac{d^2z}{dt^2}\hat{z}. \tag{7.30}$$

---

[2] At a redshift $z \sim 13$, for the cosmology fans out there.

Since an axisymmetric mass distribution has no force acting in the azimuthal direction, the azimuthal component of the acceleration is, from Equation 7.30,

$$\frac{1}{R}\frac{d}{dt}\left(R^2\frac{d\theta}{dt}\right) = 0, \qquad (7.31)$$

implying that

$$R^2\frac{d\theta}{dt} = j_z = \text{constant.} \qquad (7.32)$$

This tells us that j_z, the component of specific angular momentum associated with rotation about the z axis, is an integral of motion. The equation of motion in the R direction (combining Equations 7.29 and 7.30) is

$$\frac{d^2R}{dt^2} = -\frac{\partial\Phi}{\partial R} + R\left(\frac{d\theta}{dt}\right)^2. \qquad (7.33)$$

Using the substitution $d\theta/dt = j_z/R^2$, this equation takes the form

$$\frac{d^2R}{dt^2} = -\frac{\partial\Phi_{\text{eff}}}{\partial R}, \qquad (7.34)$$

where the effective potential is

$$\Phi_{\text{eff}}(R, z) = \Phi(R, z) + \frac{j_z^2}{2R^2}. \qquad (7.35)$$

Since the added centrifugal term $\propto R^{-2}$ in the effective potential is independent of z, we can write the equation of motion in the z direction as

$$\frac{d^2z}{dt^2} = -\frac{\partial\Phi}{\partial z} = -\frac{\partial\Phi_{\text{eff}}}{\partial z}. \qquad (7.36)$$

So far, we have simply assumed the mass distribution is axisymmetric. However, when studying galactic disks, it is often helpful to assume that the mass distribution has reflection symmetry about the midplane of the disk. That is, we can assume that there exists a $z = 0$ plane (the midplane) such that $\Phi(R, z) = \Phi(R, -z)$. For example, a potential that produces a flat rotation curve in its midplane is the logarithmic potential of Equation 6.73:

$$\Phi(R, z) = \frac{1}{2}v_0^2\ln(R^2 + z^2/q^2). \qquad (7.37)$$

When $q < 1$, this represents a potential Φ whose equipotential surfaces are oblate spheroids, and which has a constant circular speed v_0 in the $z = 0$ plane.

The effective potential for a test mass of specific angular momentum j_z in the logarithmic potential of Equation 7.37 is

$$\Phi_{\text{eff}}(R, z) = \frac{1}{2}v_0^2\ln(R^2 + z^2/q^2) + \frac{j_z^2}{2R^2}. \qquad (7.38)$$

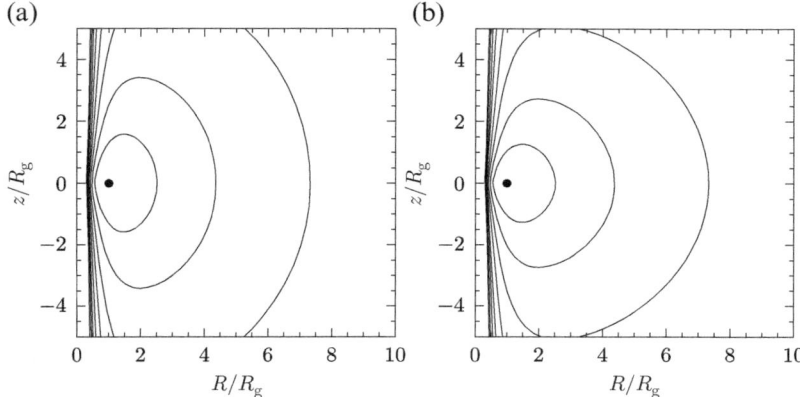

Figure 7.2 Effective potential of Equation 7.38 with equipotential contours drawn at intervals $\Delta\Phi_{\text{eff}} = 0.5v_0^2$. (a) Spherical potential, with $q = 1$. (b) Oblate potential, with $q = 0.8$. Distances are in units of $R_{\text{g}} = j_z/v_0$.

Figure 7.2 shows the effective potential of Equation 7.38 in the cases $q = 1$ and $q = 0.8$.

The effective potential climbs rapidly as $R \to 0$; the centrifugal term, proportional to j_z^2/R^2, prevents test masses with high j_z from approaching the z axis. For any potential with reflection symmetry across the $z = 0$ plane, it must be true that

$$\left.\frac{\partial\Phi_{\text{eff}}}{\partial z}\right|_{z=0} = 0. \tag{7.39}$$

By symmetry, a test mass in the $z = 0$ plane is not accelerated in the z direction. However, the only way such a mass can avoid being accelerated in the R direction is to be at the critical radius R_{g} where

$$\left.\frac{\partial\Phi_{\text{eff}}}{\partial R}\right|_{(R_{\text{g}},0)} = \left.\frac{\partial\Phi}{\partial R}\right|_{(R_{\text{g}},0)} - \frac{j_z^2}{R_{\text{g}}^3} = 0, \tag{7.40}$$

making use of Equation 7.35 for the definition of the effective potential. (As an example, the logarithmic potential of Equation 7.37 has a critical radius $R_{\text{g}} = j_z/v_0$.) Thus, a test mass that remains at coordinates $(R,z) = (R_{\text{g}},0)$ must be on a circular orbit with circular speed

$$v_{\text{circ}}^2 = \left[R\frac{\partial\Phi}{\partial R}\right]_{(R_{\text{g}},0)} = \left(\frac{j_z}{R_{\text{g}}}\right)^2 \tag{7.41}$$

and angular speed

$$\Omega^2 = \left[\frac{1}{R}\frac{\partial\Phi}{\partial R}\right]_{(R_{\text{g}},0)} = \left(\frac{j_z}{R_{\text{g}}^2}\right)^2. \tag{7.42}$$

A circular orbit minimizes the specific energy ϵ for a given value of j_z. The value of ϵ for a circular orbit in the midplane of an axisymmetric potential is

$$\epsilon_c = \frac{1}{2} v_{circ}^2 + \Phi(R_g, 0). \tag{7.43}$$

For the logarithmic potential of Equation 7.37, whose circular speed is $v_{circ} = v_0$, this becomes

$$\epsilon_c = \frac{1}{2} v_0^2 + v_0^2 \ln R_g = v_0^2 [1/2 + \ln(j_z/v_0)]. \tag{7.44}$$

If the test mass of given j_z has $\epsilon > \epsilon_c$, its orbit is not circular, and is not necessarily confined to the $z = 0$ plane. Although it must stay in the region where $\Phi_{eff}(R, z) \leq \epsilon$, it is possible to have differently shaped orbits within this region. Figure 7.3, for instance, shows two different orbits within the logarithmic effective potential plotted in Figure 7.2(b). (Although Figure 7.3 shows only R and z, the value of θ can be computed from the requirement that $d\theta/dt = j_z/R^2$.) Although the two orbits in Figure 7.3 have the same j_z and ϵ, and thus are restricted to the same region of the potential, the detailed shapes of the orbits are different.

In principle, an axisymmetric potential with $\Phi(R, z) = \Phi(R, -z)$ can host extremely non-circular orbits. In practice, real stars in real disk galaxies are often on nearly circular orbits with $z \sim 0$ and $R \sim R_g$. In this case, it is helpful to expand the effective potential as a Taylor series around its minimum at $z = 0$, $x \equiv R - R_g = 0$. Given the symmetry about $z = 0$, the terms that are odd in z vanish, and we have the expansion

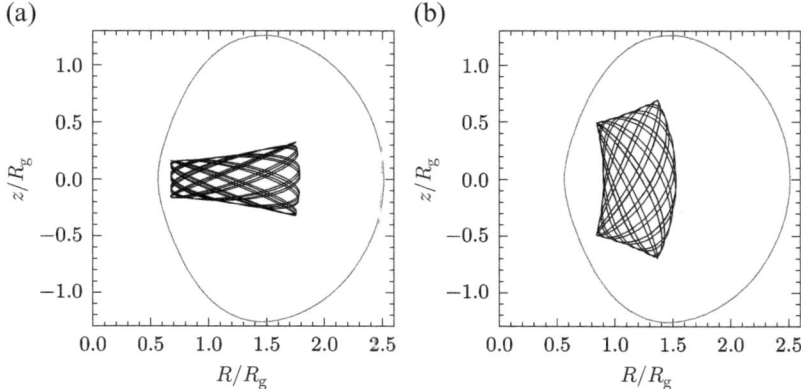

Figure 7.3 Two orbits in the effective potential of Equation 7.38, with $q = 0.8$. Both orbits have the same j_z and ϵ, with $\epsilon = \epsilon_c + v_0^2/2$. The bounding equipotential contour, where $\Phi_{eff} = \epsilon$, is shown as the closed curve. The orbits have been integrated for a time $t \sim 100 R_g/v_0 \sim 100 j_z/v_0^2$. [Calculated using galpy]

$$\Phi_{\text{eff}}(R, z) \approx \Phi_{\text{eff}}(R_{\text{g}}, 0) + \frac{1}{2}\kappa^2 x^2 + \frac{1}{2}\nu_z^2 z^2 + O(xz^2). \tag{7.45}$$

The parameter κ in Equation 7.45 is the **radial frequency**, given by the relation

$$\kappa^2 = \left.\frac{\partial^2 \Phi_{\text{eff}}}{\partial R^2}\right|_{(R_{\text{g}},0)} = \left[\frac{\partial^2 \Phi}{\partial R^2} + 3\frac{j_z^2}{R^4}\right]_{(R_{\text{g}},0)}. \tag{7.46}$$

The parameter ν_z in Equation 7.45 is the **vertical frequency**, given by the relation

$$\nu_z^2 = \left.\frac{\partial^2 \Phi_{\text{eff}}}{\partial z^2}\right|_{(R_{\text{g}},0)} = \left.\frac{\partial^2 \Phi}{\partial z^2}\right|_{(R_{\text{g}},0)}. \tag{7.47}$$

In the limit of small x and z, the equations of motion in the R and z directions are

$$\frac{d^2 x}{dt^2} = -\kappa^2 x, \qquad \frac{d^2 z}{dt^2} = -\nu_z^2 z. \tag{7.48}$$

In each direction, it's the familiar harmonic oscillator equation (as in Equation 7.17); however, the frequency κ differs from the frequency ν_z in this case. The solutions for the equations of motion are

$$x(t) = R(t) - R_{\text{g}} = x_0 \cos(\kappa t), \tag{7.49}$$

$$z(t) = z_0 \cos(\nu_z t - \Delta_z), \tag{7.50}$$

where x_0 and z_0 are positive, and $t = 0$ is chosen to be an instant when $x = x_0$. The motions in the two orthogonal directions are decoupled, with different frequencies, different amplitudes, and different phases.

The amplitudes x_0 and z_0, plus the phase shift Δ_z, are dictated by the initial conditions. The frequencies κ and ν_z, however, are determined by the potential $\Phi(R, z)$. As we found in Section 6.5, Poisson's equation for a thin disk with a flat rotation curve can be approximated as (Equation 6.166)

$$\frac{\partial^2 \Phi}{\partial z^2} = 4\pi G \rho(R, z). \tag{7.51}$$

Substituting this relation into Equation 7.47, we find that the vertical frequency is

$$\nu_z = \left[4\pi G\rho(R_{\text{g}}, 0)\right]^{1/2} = 2.38 \times 10^{-15}\,\text{s}^{-1} \left[\frac{\rho(R_{\text{g}}, 0)}{0.1\,\text{M}_\odot\,\text{pc}^{-3}}\right]^{1/2}. \tag{7.52}$$

This means that a star on a nearly circular orbit will have excursions in the z direction with period

$$\mathcal{P}_z = \frac{2\pi}{\nu_z} = 83.6\,\text{Myr} \left[\frac{\rho(R_{\text{g}}, 0)}{0.1\,\text{M}_\odot\,\text{pc}^{-3}}\right]^{-1/2}. \tag{7.53}$$

The Sun, for instance, has $v_z = 2.3 \times 10^{-15}\,\mathrm{s}^{-1}$ and $\mathcal{P}_z = 87\,\mathrm{Myr}$, assuming a local density $\rho = 0.093\,\mathrm{M}_\odot\,\mathrm{pc}^{-3}$, from Equation 6.169. By comparison, the orbital period of the Sun about the galactic center is $\mathcal{P}_\theta = 214\,\mathrm{Myr} \approx 2.46 \mathcal{P}_z$, assuming a nearly circular orbit with speed $v_{\mathrm{circ}} = 235\,\mathrm{km\,s}^{-1}$ and radius $R = 8.2\,\mathrm{kpc}$. Since the vertical period $\mathcal{P}_z$ depends on the local density ρ, while the orbital period $\mathcal{P}_\theta$ depends on an integral of ρ over the entire mass distribution, we don't expect $\mathcal{P}_\theta/\mathcal{P}_z$ to be a rational number.

The Sun passes through the midplane of our galaxy once every $\sim$43 Myr; it has undergone such a passage relatively recently. The Sun is currently at a distance $z \approx 21\,\mathrm{pc}$ north of the midplane and is traveling northward with $dz/dt \approx 8\,\mathrm{km\,s}^{-1}$. Using Equation 7.50, we can write

$$z(t) = z_0 \cos(v_z t - \Delta_z) \approx 21\,\mathrm{pc} = 6.5 \times 10^{14}\,\mathrm{km} \tag{7.54}$$

and

$$\frac{dz}{dt} = -v_z z_0 \sin(v_z t - \Delta_z) = 8\,\mathrm{km\,s}^{-1}. \tag{7.55}$$

Combining Equations 7.54 and 7.55, we find that

$$\tan(v_z t - \Delta_z) = -\frac{dz/dt}{v_z z} = -5.4, \tag{7.56}$$

assuming $v_z = 2.3 \times 10^{-15}\,\mathrm{s}^{-1}$. The appropriate solution, given the signs of z and dz/dt, is $v_z t - \Delta_z = -1.39 = -0.44\pi$. The time elapsed since the most recent midplane passage, at $v_z t - \Delta_z = -\pi/2$, is then

$$\Delta t = \frac{\pi/2 - 1.39}{v_z} = 7.9 \times 10^{13}\,\mathrm{s} = 2.5\,\mathrm{Myr}. \tag{7.57}$$

From Equation 7.54, we find that the Sun's maximum excursion from the midplane is $z_0 = 110\,\mathrm{pc}$; since this is smaller than the disk scale height $h_z \approx 200\,\mathrm{pc}$, the assumption of uniform ρ in Equation 7.52 is not a horrifically bad one.

Motion in the (R, θ) plane for nearly circular orbits in an axisymmetric potential is more interesting than the simple harmonic motion in the z direction. To make sense of motion in the (R, θ) plane, adopt a reference frame rotating about the z axis with an angular speed $\Omega = v_{\mathrm{circ}}/R_{\mathrm{g}} = j_z/R_{\mathrm{g}}^2$. In this frame, we choose x and y axes as illustrated in Figure 7.4, with $(x, y) = (0, 0)$ corresponding to the position of a particle on a perfectly circular orbit with radius R_{g} and angular momentum j_z. If we take a test particle on a circular orbit of radius R_{g} and slightly perturb it in the $z = 0$ plane, then the point $(x, y) = (0, 0)$ is the **guiding center** for the motion of the particle in the rotating frame. In this frame, the particle undergoes harmonic oscillation in the radial direction, with (Equation 7.49)

$$x(t) = x_0 \cos(\kappa t), \tag{7.58}$$

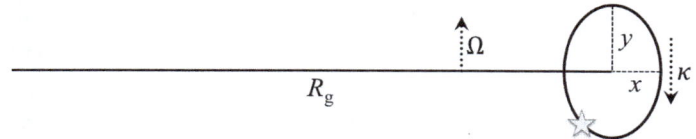

Figure 7.4 In a frame rotating with angular speed $\Omega = j_z/R_g^2$, a test mass moves at angular frequency κ around an elliptical epicycle.

where κ is the radial frequency. Since the test mass alternately moves closer to and farther from the potential center, conservation of j_z means its orbital speed must alternately increase and decrease. In the inertial frame, the angular speed of the particle about the potential center is

$$\frac{d\theta}{dt} = \frac{j_z}{(R_g + x)^2} \approx \Omega\left[1 - \frac{2x_0}{R_g}\cos(\kappa t)\right], \tag{7.59}$$

given $\Omega = j_z/R_g^2$ and assuming $x_0 \ll R_g$. In the rotating frame, this implies that the physical speed of the particle perpendicular to the x axis is

$$\frac{dy}{dt} = R\frac{d\theta}{dt} - R\Omega \approx -2\Omega x_0 \cos(\kappa t). \tag{7.60}$$

Integration yields

$$y(t) = -\frac{2\Omega}{\kappa}x_0 \sin(\kappa t). \tag{7.61}$$

The particle thus undergoes **epicyclic motion** around a small ellipse whose axis ratio is $x_0/y_0 = \kappa/(2\Omega)$.

Since j_z is conserved, harmonic motion in the R direction must result in motion on a small epicycle in the (x, y) plane. Thus, the radial frequency κ is also known as the **epicycle frequency**.[3] Conservation of j_z also dictates that the sense of motion of the test mass on its epicycle (clockwise in Figure 7.4) is opposite to that of the guiding center on its circular orbit. In the equation for the epicycle frequency κ (Equation 7.46), we can use the relation for the angular speed Ω (Equation 7.42) to make the substitutions

$$\frac{\partial^2 \Phi}{\partial R^2} = \frac{\partial}{\partial R}(R\Omega^2) \quad \text{and} \quad \frac{j_z^2}{R^4} = \Omega^2. \tag{7.62}$$

As a result, we can write κ in terms of Ω, with

$$\kappa^2 = \left[\frac{\partial}{\partial R}(R\Omega^2) + 3\Omega^2\right]_{(R_g,0)} = 4\Omega^2\left[1 + \frac{1}{2}\frac{\partial \ln \Omega}{\partial \ln R}\right]_{(R_g,0)}. \tag{7.63}$$

---

[3] Epicycles have received a bad press ever since Kepler discarded them for planetary motion. However, in the case of nearly circular orbits, the epicycle approximation is actually a useful mathematical tool for describing motion.

The ratio κ/Ω thus depends on the shape of the potential in which the test mass is orbiting. In a potential with a flat rotation curve, appropriate for a disk galaxy, $\Omega = v_{\mathrm{circ}}/R \propto R^{-1}$, and $\kappa/\Omega = \sqrt{2}$. This means, assuming the Sun has an orbital period $\mathcal{P}_\theta = 2\pi/\Omega = 214\,\mathrm{Myr}$, its epicycle period is $\mathcal{P}_\kappa \approx \mathcal{P}_\theta/\sqrt{2} \approx 150\,\mathrm{Myr}$. The axis ratio of an epicycle in a potential with a flat rotation curve is $x_0/y_0 = 1/\sqrt{2}$.

We can use the Sun as our example of a test mass undergoing epicyclic motion, just as we used it as our example for motion in the z direction. The Sun is currently moving toward the galactic center, with $dx/dt \approx -11\,\mathrm{km\,s^{-1}}$. It is also moving faster than its guiding center in the direction of galactic rotation, with $dy/dt \approx 12\,\mathrm{km\,s^{-1}}$.[4] Using Equation 7.58, we can write

$$\frac{dx}{dt} = -\kappa x_0 \sin(\kappa t) = -11\,\mathrm{km\,s^{-1}}. \tag{7.64}$$

Using Equation 7.61, we can also write

$$\frac{dy}{dt} = -\sqrt{2}\kappa x_0 \cos(\kappa t) = 12\,\mathrm{km\,s^{-1}}, \tag{7.65}$$

assuming a flat rotation curve. Combining Equations 7.64 and 7.65, we find

$$\tan(\kappa t) = \sqrt{2}\frac{dx/dt}{dy/dt} = -1.30. \tag{7.66}$$

The appropriate solution, given the signs of dx/dt and dy/dt, is $\kappa t = 2.23 = 0.71\pi$. From Equation 7.64, this implies a maximum excursion in the radial direction of

$$x_0 = -\frac{dx/dt}{\kappa \sin(0.71\pi)} \approx 340\,\mathrm{pc}. \tag{7.67}$$

The maximum excursion in the azimuthal direction is then $y_0 = \sqrt{2}x_0 \approx 480\,\mathrm{pc}$. Since $x_0 \approx 0.04R_{\mathrm{g}}$, the epicyclic approximation is appropriate for the Sun's orbit.

7.3 Orbits in Non-axisymmetric Potentials

Not every stellar system is adequately approximated as being spherical, or even as being axisymmetric. It is time, therefore, to make a few comments about non-axisymmetric potentials, and the orbits that they support. As a simple case, consider a two-dimensional potential $\Phi(x, y)$; this might represent the potential felt by particles orbiting in a thin non-circular disk. For concreteness, consider the potential

[4] These speeds in the x and y direction imply that the Sun is approximately at the location of the five-pointed star in Figure 7.4; closer to the galactic center than average, but lagging behind the guiding center in the azimuthal direction.

$$\Phi(x, y) = \frac{1}{2}v_0^2 \ln\left(1 + \frac{x^2 + y^2/q^2}{r_c^2}\right), \tag{7.68}$$

where $q \leq 1$. The parameter q is the axis ratio of the equipotential contours; in the limit $q = 1$, this potential leads to a flat rotation curve, with $v_{circ} = v_0$, outside a core of radius r_c. Within the central core, the potential can be expanded as

$$\Phi(x, y) \approx \frac{v_0^2}{2r_c^2}(x^2 + y^2/q^2). \tag{7.69}$$

The equations of motion within the central core are thus

$$\frac{d^2x}{dt^2} = -\omega_x^2 x, \qquad \frac{d^2y}{dt^2} = -\omega_y^2 y, \tag{7.70}$$

with $\omega_x = v_0/r_c$ and $\omega_y = \omega_x/q$. Once again, it's the familiar harmonic oscillator, as in Equations 7.17 and 7.48. The solution for this oscillator is

$$x(t) = x_0 \cos(\omega_x t), \tag{7.71}$$
$$y(t) = y_0 \cos(\omega_y t - \delta), \tag{7.72}$$

where the amplitudes x_0 and y_0, as well as the phase shift δ, are set by the initial conditions. If $q = 1$, then $\omega_x = \omega_y$, and the orbit is an ellipse centered on the origin (as shown in Figure 7.1, for instance). However, if $q < 1$, then $\omega_x \neq \omega_y$, and the orbit is a **Lissajous curve**. Figure 7.5 shows some Lissajous curves for which $\omega_x/\omega_y = q$ is a rational number. In general, however, we don't expect q to be rational; for irrational values of q, a Lissajous curve is not closed.

In the context of stellar dynamics, a Lissajous curve is an example of a **box orbit**. The angular momentum of a test mass on a box orbit is not conserved. If $\omega_x/\omega_y = q$ is irrational, then after some finite time the test mass will come arbitrarily close to the center of the potential. Box orbits are generally found in non-axisymmetric potentials that have well-defined cores with nearly constant mass density. Outside the constant-density core, motion becomes more complicated than a simple harmonic oscillator. However, numerical simulations tell us that there are two main families of orbits in non-axisymmetric potentials. First are the box orbits, which have no unique sense of orbital motion, and which bring stars arbitrarily close to the center of the potential. Next are the **loop orbits**, which have a unique sense of orbital motion (although j_z is not strictly conserved if the potential is not axisymmetric). Stars on loop orbits stay away from the center of the potential.

Figure 7.6 shows examples of two-dimensional box and loop orbits in the logarithmic potential of Equation 7.68. Both orbits have specific energy $\epsilon = 1.5v_0^2$, and thus have the same bounding equipotential contour, where $\Phi(x, y) = \epsilon$. This bounding contour is shown as the ellipse in each panel of Figure 7.6.

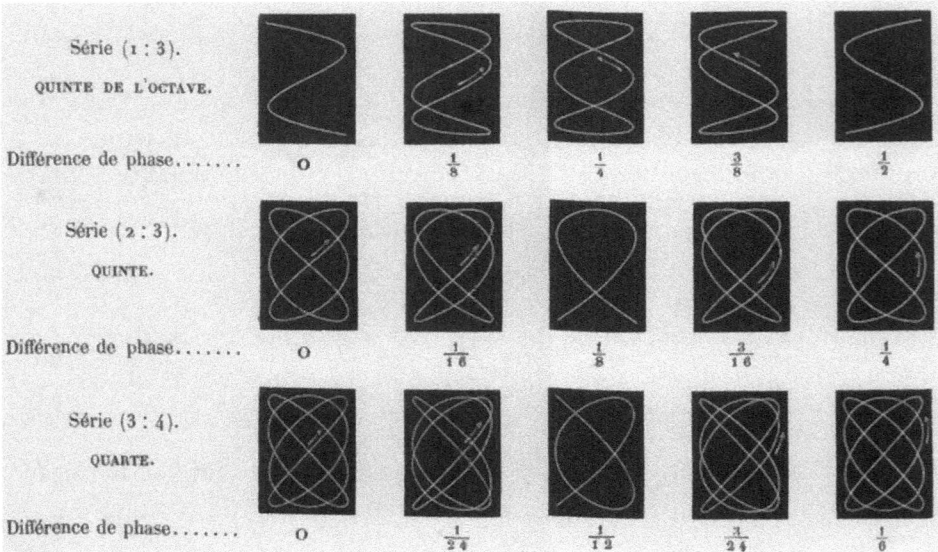

Figure 7.5 Lissajous curves, as traced by Jules Lissajous by reflecting light beams from vibrating mirrors. The three rows, from top to bottom, represent $\omega_x/\omega_y = 3/1$, $3/2$, and $4/3$. The phase shift, beneath each individual image, is given as a fraction of 2π. [Lissajous 1857]

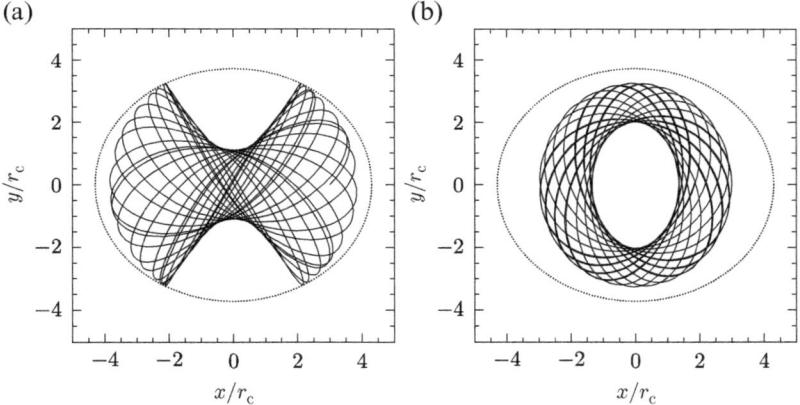

Figure 7.6 A box orbit (a) and loop orbit (b) of identical specific energy $\epsilon \approx 1.5v_0^2$ in the non-axisymmetric potential of Equation 7.68; an axis ratio $q = \sqrt{3}/2$ is assumed. The bounding equipotential contour is shown as the dotted ellipse. The orbits have been integrated for $t \sim 300r_c/v_0$. [Calculated using galpy]

The two-dimensional elliptical potential of Equation 7.68 has a three-dimensional ellipsoidal equivalent:

$$\Phi(x, y, z) = \frac{1}{2}v_0^2 \ln\left(1 + \frac{x^2 + y^2/\beta^2 + z^2/\gamma^2}{r_c^2}\right), \qquad (7.73)$$

where the primary axes are chosen so that $1 \geq \beta \geq \gamma > 0$. The limiting case $\beta = 1$ corresponds to oblate spheroids, while the case $\beta = \gamma$ corresponds to prolate spheroids. In general, however, $1 > \beta > \gamma$, and the equipotential surfaces are **triaxial ellipsoids**. In the central core, the potential is approximately (compare to Equation 7.69)

$$\Phi(x, y, z) \approx \frac{v_0^2}{2r_c^2}(x^2 + y^2/\beta^2 + z^2/\gamma^2). \tag{7.74}$$

Along each primary axis, the motion is that of a harmonic oscillator, with $\omega_x = v_0/r_c$, $\omega_y = \omega_x/\beta$, and $\omega_z = \omega_x/\gamma$. Thus, orbits confined to the core are three-dimensional Lissajous orbits.

Outside the central core, the triaxial potential of Equation 7.73 supports both box orbits and **tube orbits**, the three-dimensional equivalent of loop orbits. Stars on short-axis tube orbits maintain a constant sense of direction around the short axis of the potential; stars on long-axis tube orbits maintain a constant sense of direction around the long axis of the potential. There are no stable tube orbits around the intermediate axis of a triaxial potential. It is possible to build a self-consistent triaxial galaxy by adding together stars on box orbits, short-axis tube orbits, and long-axis tube orbits. In general, galaxies that are nearly oblate will contain a high percentage of short-axis tube orbits. Galaxies that are highly triaxial will contain a high percentage of box orbits.

Determining the three-dimensional shape of an elliptical galaxy is challenging, given that we can see only its projected two-dimensional surface brightness. However, statistical studies of large populations of galaxies reveal that elliptical galaxies are usually nearly oblate, with typical axis ratios $1 : 0.98 : 0.70$. The outer regions of an elliptical galaxy are typically more flattened (with a smaller value of γ) than its central region. This is because the shape of the stellar distribution is affected by the presence of the galaxy's central supermassive black hole. More massive elliptical galaxies have larger stellar velocity dispersions and, in general, more massive central black holes. The observationally determined correlation between the central black hole mass m_{bh} and the line-of-sight stellar velocity dispersion σ is

$$m_{bh} \approx 3 \times 10^8 \, M_\odot \left(\frac{\sigma}{200 \, \mathrm{km \, s^{-1}}} \right)^{4.4}. \tag{7.75}$$

A star on a box orbit will eventually come close enough to the central black hole to have a close gravitational encounter, and be kicked in a random direction. From Equation 6.9, the resulting critical impact parameter for a close gravitational encounter is

$$b_{cge} \sim \frac{2Gm_{bh}}{v^2}, \tag{7.76}$$

where v is the star's speed. Assuming $v \sim \sigma$, and using the m_{bh}–σ relation of Equation 7.75, this becomes

$$b_{cge} \sim 60\,pc \left(\frac{\sigma}{200\,km\,s^{-1}}\right)^{2.4} \sim 120\,pc \left(\frac{m_{bh}}{10^9\,M_\odot}\right)^{0.55}. \qquad (7.77)$$

The central black hole, as it busts up box orbits, converts them to more nearly circular chaotic orbits. As a star shuttles back and forth on its box orbit, it typically takes a time $\sim 100 t_{dyn}$ before it has a close gravitational encounter with the central black hole. If the age of the elliptical galaxy is t_{sys}, then its central regions, where $t_{dyn} \ll 0.01 t_{sys}$, are nearly spherical. The outer regions, where the central black hole has not had time to significantly influence the box orbits, can still be triaxial.

A star on a purely radial orbit will have a more dramatic interaction with the central black hole: it will be **tidally disrupted**. As a star with mass $m_\star$ and radius $R_\star$ approaches a black hole with mass m_{bh}, it will experience tidal distortions with height (Equation 4.16)

$$h \sim h_T = \frac{3 m_{bh}}{2 m_\star} \left(\frac{R_\star}{r}\right)^3 R_\star, \qquad (7.78)$$

where r is the distance of the star from the black hole. The star will be torn apart (or "tidally disrupted," to use the polite term) when $h \sim R_\star$. This occurs at the tidal disruption radius,

$$r_{td} \sim \left(\frac{3 m_{bh}}{2 m_\star}\right)^{1/3} R_\star \sim \left(\frac{9 m_{bh}}{8 \pi \rho_\star}\right)^{1/3}, \qquad (7.79)$$

where $\rho_\star$ is the mean density of the star. Scaled to a $10^9\,M_\odot$ black hole and a Sun-like star, this becomes

$$r_{td} \sim 5\,au \left(\frac{m_{bh}}{10^9\,M_\odot}\right)^{1/3} \left(\frac{\rho_\star}{1.4\,g\,cm^{-3}}\right)^{-1/3}, \qquad (7.80)$$

where 1 au is equal to 4.8 μpc. Since $r_{td} \ll b_{cge}$, tidal disruption of stars will be very rare compared to close gravitational encounters of stars with the central black hole.

Exercises

7.1 Imagine what would happen if the gravitational potential of a point mass m were a short-range *Yukawa potential*. Such a potential has the form

$$\Phi(r) = -\frac{Gm}{r} \exp(-r/\lambda), \qquad (7.81)$$

where λ is the screening length of the potential.

(a) What would be the circular speed v_{circ} in this potential?

(b) A nearly circular orbit can be described in the epicyclic approximation. For the potential of Equation 7.81, what is the epicycle frequency $\kappa(r)$? What will be the fate of a test mass on a nearly circular orbit with radius $r \gg \lambda$?

(c) If Newtonian gravity did have a Yukawa potential, what observations place the largest lower limit on the screening length λ?

7.2 A steady-state axisymmetric system has reflection symmetry across the $z = 0$ plane, so that $\Phi(R, z) = \Phi(R, -z)$.

(a) Show that the radial frequency κ and the vertical frequency v_z are related by the expression

$$\kappa^2 + v_z^2 = 4\pi G\rho(R, 0) + 2\Omega^2, \qquad (7.82)$$

where $\rho(R, 0)$ is the density in the $z = 0$ plane and $\Omega(R)$ is the angular speed on a circular orbit in the $z = 0$ plane.

(b) Suppose that the mass distribution has finite extent, so that $\rho(R, 0) = 0$ outside the truncation radius R_t. What is $\kappa(R)$ for a test mass outside the truncation radius? What is $v_z(R)$ for the same test mass?

7.3 One potential with interesting analytic properties is the **isochrone** potential,

$$\Phi_{\text{iso}}(r) = -\frac{GM}{b + (b^2 + r^2)^{1/2}}, \qquad (7.83)$$

whose properties were studied by Michel Hénon in 1959.

(a) Show that the density ρ_{iso} that gives rise to the isochrone value has a finite value ρ_c at $r = 0$. Compute the value of ρ_c in terms of M and b.

(b) What is the dependence of ρ_{iso} on r in the limit $r \gg b$?

(c) Hénon used the adjective "isochrone" for this potential because orbits with the same value of ϵ have the same radial period $\mathcal{P}_r$ in this potential, regardless of j. Using the (relatively simple) example of a radial orbit with $j = 0$, compute $\mathcal{P}_r$ as a function of the specific energy ϵ of a test mass.

7.4 A two-body system has a standard gravitational parameter $\mu = G(m_1 + m_2)$ and a mean separation a between the two objects. In the limit $\mu/a \ll c^2$, the effects of general relativity can be mimicked by slightly altering the Keplerian potential, to

$$\Phi(r) = \Phi_K(r) + \Phi_{\text{GR}}(r) = -\frac{\mu}{r} - \frac{\mu j^2}{c^2 r^3}, \qquad (7.84)$$

where j is the specific orbital angular momentum.

(a) For nearly circular orbits, compute the angular speed $\Omega(r)$ and epicycle frequency $\kappa(r)$ in the altered potential.

(b) Show that during one orbital period, $\mathcal{P} = 2\pi/\Omega$, the periapsis precesses through an angle

$$\delta\theta_{\mathrm{pe}}(r) \propto \frac{\mu}{c^2 r}. \tag{7.85}$$

If $\delta\theta_{\mathrm{pe}}$ is in radians, what is the normalization of Equation 7.85?

(c) For a two-body system consisting of Mercury and the Sun, what rate (in arcseconds per century) would you calculate for the precession of the perihelion of Mercury due to general relativistic effects?

8

Collisionless Stellar Systems

Sir James Jeans
Always says what he means;
He is really perfectly serious
About the Universe being Mysterious.

E. Clerihew Bentley (1875–1956)
Punch, vol. 196, issue 5100, p. 39 [1939 Jan 11]

We have defined stellar dynamics as the study of systems that contain many objects (which can usually be approximated as point masses) that interact via Newtonian gravity. The name "stellar" dynamics implies that the gravitationally interacting objects are stars. However, we saw in Section 7.3 that the techniques of stellar dynamics can be applied to the interaction of stars with black holes. In addition, since the techniques of stellar dynamics don't require the interacting objects to have a particular mass, they can be applied to the dark matter in a galactic halo as well as to the garishly glowing stars of the galaxy.

It is appropriate that the techniques of stellar dynamics can be applied to non-stellar objects, since many of these techniques were drawn from outside the realm of astronomy. The virial theorem, as we saw in Section 6.1, was proposed by Clausius as an extremely general law. However, its first applications were in the field of thermodynamics. Similarly, the collisionless Boltzmann equation (Equation 6.50) was first derived by Ludwig Boltzmann in the context of fluid dynamics. However, the astronomical community can take pride in the fact that some tools of stellar dynamics, such as the Jeans equations, were actually developed specifically for astronomical purposes.

8.1 Distribution Functions and the Jeans Theorem

Sir James Jeans was a productive man. In addition to developing the Jeans equations and writing popular books such as *The Mysterious Universe*, he also formulated what is now known as the **Jeans theorem**. To grasp the Jeans theorem, and its importance for stellar dynamics, start by recalling that an "integral of motion" is a function $I(\vec{r}, \vec{v})$ that is constant as a star moves along its orbit. Examples of an integral of motion are the specific energy ϵ in a static potential, and the specific angular momentum $\vec{j}$ in a static spherical potential. By definition,

$$\frac{d}{dt} I(\vec{r}(t), \vec{v}(t)) = 0. \tag{8.1}$$

This can be written more explicitly as

$$\vec{\nabla} I \cdot \frac{d\vec{r}}{dt} + \frac{\partial I}{\partial \vec{v}} \cdot \frac{d\vec{v}}{dt} = 0. \tag{8.2}$$

However, if the only force is that of gravity, dictated by a potential $\Phi(\vec{r})$, this becomes

$$\vec{\nabla} I \cdot \vec{v} - \frac{\partial I}{\partial \vec{v}} \cdot \vec{\nabla}\Phi = 0. \tag{8.3}$$

This should look familiar: it is identical to the collisionless Boltzmann equation (Equation 6.50) for a steady-state system. Thus, in a system that has reached a steady state, the distribution function $f(\vec{r}, \vec{v})$ is an integral of motion.

The Jeans theorem states that if a potential has n integrals of motion, I_1 through I_n, then *any* function of the integrals of motion yields a steady-state solution of the collisionless Boltzmann equation. To see why this is true, consider an arbitrary function $f(I_1, I_2, \ldots, I_n)$ of the integrals of motion. Then

$$\frac{df}{dt} = \sum_{j=1}^{n} \frac{\partial f}{\partial I_j} \frac{dI_j}{dt}. \tag{8.4}$$

However, since $dI_j/dt = 0$ (that's part of the definition of an integral of motion),

$$\frac{df}{dt} = 0 = \vec{\nabla} f \cdot \frac{d\vec{r}}{dt} + \frac{\partial f}{\partial \vec{v}} \cdot \frac{\partial \vec{v}}{\partial t}. \tag{8.5}$$

However, when the acceleration results purely from Newtonian gravity, this relation becomes

$$0 = \vec{\nabla} f \cdot \vec{v} - \frac{\partial f}{\partial \vec{v}} \cdot \vec{\nabla}\Phi. \tag{8.6}$$

This is, once again, the collisionless Boltzmann equation for a steady-state system.

Although any function of the integrals of motion is a mathematically valid solution for the collisionless Boltzmann equation, not every function is a *physically* valid distribution function for a stellar system. For instance, physical reality dictates that $f(\vec{r}, \vec{v}) \geq 0$ over all phase space; there cannot be a negative number of stars at a given location with a given velocity. To see what functions might be valid distribution functions, start with a system that has a spherical potential, $\Phi(r)$. For such a system, the specific energy $\epsilon = v^2/2 + \Phi$ and the three components of the specific angular momentum $\vec{j}$ are integrals of motion. To simplify things still further, let's assume the stellar system has no net rotation; with no preferred axis of rotation, we expect that the distribution function depends only on the magnitude of $\vec{j}$, and not on its direction. Thus, we expect a distribution function $f(\epsilon, j)$ to correspond to a non-rotating, spherical, steady-state system. In one final burst of simplicity, we will attempt a distribution function of the form $f(\epsilon)$, and see what type of spherical stellar system it produces.

8.1.1 Spherical Systems

In a bound stellar system, every star has a negative specific energy ϵ. The most loosely bound star in the system (the one with ϵ closest to zero) has a specific energy ϵ_0. For ease of calculation, it is useful to define a *relative* specific energy $\mathcal{E} \equiv \epsilon_0 - \epsilon$, which is non-negative for every star in the system. The relative specific energy can also be written as $\mathcal{E} = \Psi(r) - v^2/2$, where Ψ is the *relative* potential, $\Psi(r) \equiv \epsilon_0 - \Phi(r)$. In terms of the relative specific energy, the distribution function is

$$f(\mathcal{E}) \geq 0 \quad \text{when } \mathcal{E} > 0, \tag{8.7}$$

$$f(\mathcal{E}) = 0 \quad \text{when } \mathcal{E} \leq 0. \tag{8.8}$$

Given $f = f(\mathcal{E}) = f(\Psi(r) - v^2/2)$, the number density of stars in the spherical system is (Equation 6.51)

$$n_\star(r) = \int f(\Psi(r) - v_x^2/2 - v_y^2/2 - v_z^2/2)\, dv_x\, dv_y\, dv_z. \tag{8.9}$$

The mean square velocity of the stars in any direction is (Equation 6.60)

$$\langle v_x^2 \rangle = \frac{1}{n_\star(r)} \int v_x^2 f(\Psi - v_x^2/2 - v_y^2/2 - v_z^2/2)\, dv_x\, dv_y\, dv_z. \tag{8.10}$$

The form of Equation 8.10 requires that $\langle v_x^2 \rangle = \langle v_y^2 \rangle = \langle v_z^2 \rangle$. If f is a function only of ϵ, then the velocity dispersion of a stellar system must be isotropic at every point. Conversely, if a stellar system is found to have a dispersion that is anisotropic, then its distribution function cannot depend solely on ϵ.

As an example, consider a distribution function that is exponential in energy:

$$f(\mathcal{E}) = \frac{n_0}{(2\pi\sigma^2)^{3/2}} \exp\left(\frac{\mathcal{E}}{\sigma^2}\right) \tag{8.11}$$

when $\mathcal{E} > 0$, and $f = 0$ otherwise. In Equation 8.11, the parameters σ^2 (with dimensions of specific energy) and n_0 (with dimensions of number density) give the correct dimensionality and normalization.

Since stars with larger $\mathcal{E}$ are more tightly bound, the exponential distribution of Equation 8.11 leads us to expect a centrally concentrated stellar distribution. But how concentrated? Written more explicitly, the exponential distribution of Equation 8.11 is

$$f(\mathcal{E}) = \frac{n_0}{(2\pi\sigma^2)^{3/2}} \exp\left(\frac{\Psi(r)}{\sigma^2} - \frac{v^2}{2\sigma^2}\right). \tag{8.12}$$

Integrated over velocity (Equation 8.9), this distribution function yields a density profile

$$n_\star(r) = n_0 \exp\left(\frac{\Psi(r)}{\sigma^2}\right). \tag{8.13}$$

Thus, from Equation 8.12, we can write the exponential distribution function as $f(\mathcal{E}) = n_\star(r)F(v)$, where

$$F(v)d^3v = \frac{1}{(2\pi\sigma^2)^{3/2}} \exp\left(-\frac{v^2}{2\sigma^2}\right) d^3v. \tag{8.14}$$

This represents an isotropic **Maxwellian distribution**, which can also be written as

$$\begin{aligned}
f_M(v)dv &= \frac{1}{(2\pi\sigma^2)^{3/2}} \exp\left(-\frac{v^2}{2\sigma^2}\right) 4\pi v^2 dv \\
&= \left(\frac{2}{\pi}\right)^{1/2} \frac{v^2}{\sigma^2} \exp\left(-\frac{v^2}{2\sigma^2}\right) \frac{dv}{\sigma}.
\end{aligned} \tag{8.15}$$

In a gas of molecules or atoms, $\sigma^2 = kT/m$, where m is the mean molecular mass. In a stellar system, $\sigma^2 = \langle v_x^2 \rangle = \langle v_y^2 \rangle = \langle v_z^2 \rangle$.

From Equation 8.13, we know that the stellar density $n_\star(r)$ and the relative potential $\Psi(r)$ have the relation

$$\Psi(r) = \sigma^2 \ln[n_\star(r)/n_0]. \tag{8.16}$$

However, we also know Poisson's equation for a spherical system,[1]

$$\frac{1}{r^2}\frac{d}{dr}\left(r^2\frac{d\Psi}{dr}\right) = -4\pi G\rho(r), \tag{8.17}$$

where $\rho(r)$ is the mass density. In our usual quest for simplicity, let's assume a stellar system in which all the mass is provided by stars of mass $m_\star$, so that

---

[1] Remember that $\Psi = \epsilon_0 - \Phi$; hence the change in sign in Equation 8.17 from the usual form of Poisson's equation.

$\rho(r) = m_\star n_\star(r)$. In this case, Equations 8.16 and 8.17 can be combined into a single differential equation for $n_\star$:

$$\frac{d}{dr}\left(r\frac{d\ln n_\star}{d\ln r}\right) = -\left[4\pi\frac{Gm_\star}{\sigma^2}\right]r^2 n_\star. \tag{8.18}$$

When in doubt, try a power law. Assuming $n_\star = Cr^{-b}$, we find $b = 2$ and $C = \sigma^2/(2\pi Gm_\star)$. Thus, a stellar system whose distribution function f is exponential in $\mathcal{E}$ turns out to be a singular isothermal sphere, with

$$\rho(r) = m_\star n_\star(r) = \frac{\sigma^2}{2\pi G}r^{-2}, \tag{8.19}$$

$$F(v) = \frac{1}{(2\pi\sigma^2)^{3/2}}\exp\left(-\frac{v^2}{2\sigma^2}\right). \tag{8.20}$$

As mentioned in Section 6.4.1, such a singular isothermal sphere can be a useful approximation to the dark halo surrounding a disk galaxy with a flat rotation curve ($v_{\text{circ}} = \sqrt{2}\sigma$).

Applying the singular isothermal sphere to actual stellar systems does have some problems to overcome. For instance, many observed stellar systems have a constant-density core (as in a Plummer profile) or a central cusp shallower than r^{-2} (as in a Hernquist profile). Another problem with isothermal spheres is that the enclosed mass M goes to infinity as $r \to \infty$. Thus, an outer truncation radius must be imposed on the stellar system in a physically realistic way. (In Section 6.4.1, we simply dictated an abrupt drop to $n_\star = 0$ at a truncation radius r_t; this was computationally useful, but not realistic.) One way to avoid having stars at large radii, corresponding to small values of $\mathcal{E}$, is to change the exponential distribution function of Equation 8.11 into that of a **King distribution**:[2]

$$f_K(\mathcal{E}) = \frac{n_0}{(2\pi\sigma^2)^{3/2}}\left[\exp(\mathcal{E}/\sigma^2) - 1\right]. \tag{8.21}$$

We can integrate the King distribution over all velocities to find the spatial density distribution n_K. To show the density of a **King model** in its most compact form, it is useful to define a *dimensionless* relative potential,

$$W(r) \equiv \frac{\Psi(r)}{\sigma^2} = \frac{\epsilon_0}{\sigma^2} - \frac{\Phi(r)}{\sigma^2}. \tag{8.22}$$

This potential has a maximum value $W_0 = \Psi(0)/\sigma^2$ at the center of the stellar distribution and goes to $W(r_t) = 0$ at an outer truncation radius r_t. This truncation radius represents the radius at which the most loosely bound star

---

[2] The "King" distribution function is no more regal than other functions; instead, it is named after Ivan King, who used it in 1966 to describe star clusters.

in the system would have $v = 0$ if it were on a radial orbit. In terms of the potential $W(r)$, the stellar density of a King model can be written as

$$n_K(r) = \frac{4}{3\sqrt{\pi}} n_0 \exp(W)\gamma(5/2, W), \tag{8.23}$$

where $\gamma(5/2, x)$ is the incomplete gamma function of order 5/2. We can use the properties of incomplete gamma functions to find n_K in the limit of small and large values of $W(r)$. When $W(r) \ll 1$, corresponding to loosely bound stars near the edge of the stellar system,

$$n_K(r) \approx \frac{8}{15\sqrt{\pi}} n_0 W^{5/2}. \tag{8.24}$$

In the opposite limit of tightly bound stars with $W(r) \gg 1$, the density is

$$n_K(r) \approx n_0 \exp(W), \tag{8.25}$$

indistinguishable from the distribution for a perfect isothermal sphere (Equation 8.13).

To find a solution for the dimensionless potential $W(r)$, we write Poisson's equation (Equation 8.17) in the form

$$\frac{1}{r^2}\frac{d}{dr}\left(r^2\frac{dW}{dr}\right) = -4\pi\frac{Gm_\star}{\sigma^2}n_K \tag{8.26}$$

$$= -\frac{16\sqrt{\pi}}{3}\frac{Gm_\star n_0}{\sigma^2}\exp[W(r)]\gamma(5/2, W(r)). \tag{8.27}$$

Equation 8.27 is a second-order differential equation that can be integrated starting at $r = 0$. This requires choosing a value for the central potential, $W_0 = \Psi(0)/\sigma^2$, and a value for dW/dr at $r = 0$. To prevent a cuspy potential, the boundary condition $dW/dr = 0$ is usually chosen. At small r, this yields a potential of the form

$$W(r) \approx W_0 - \frac{3}{2}\frac{r^2}{r_K^2}, \tag{8.28}$$

where r_K is the **King radius**.[3] Using the approximation of Equation 8.28 in Poisson's equation (Equation 8.27) leads to

$$r_K = \left(\frac{9\sigma^2}{4\pi G\rho(0)}\right)^{1/2}, \tag{8.29}$$

where $\rho(0) \approx m_\star n_0 \exp(W_0)$ is the central mass density.

The integration of Equation 8.27 can be continued numerically, stepping outward in r until the relative potential reaches $W = 0$; this gives the location of the truncation radius r_t. Once the function $W(r)$ is computed from $r = 0$ to

---

[3] The factor of 3/2 in Equation 8.28 might seem arbitrary; it was actually chosen by Ivan King so that r_K would be close to the radius at which the projected surface density drops to half its central value.

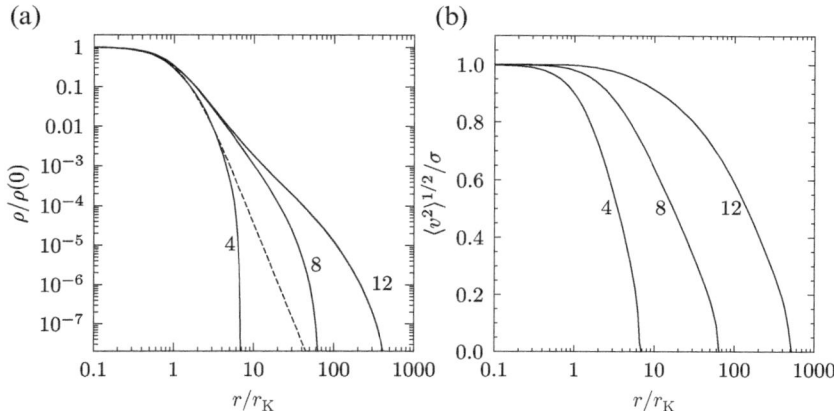

Figure 8.1 (a) Density profiles of King models with potential depth $W_0 = 4$, 8, and 12. A Plummer model with $r_c = (5/3)^{1/2} r_K$ is shown as the dashed line. (b) One-dimensional velocity dispersion $\langle v^2 \rangle^{1/2}$ for the King models shown in panel (a). [Calculated using BHKing calculator, Cosmic-Lab project]

$r = r_t$, substitution into Equation 8.23 yields the desired King model, $n_K(r)$. Figure 8.1(a) shows the density profiles for three King models, with $W_0 = 4$, 8, and 12. The larger the value of W_0, the larger the truncation radius r_t is relative to the King radius r_K. Because of this relation, King models are often described by a concentration index, $c = \log(r_t/r_K)$, rather than by W_0. For the models shown in Figure 8.1, the concentration index ranges from $c = 0.84$ when $W_0 = 4$ to $c = 2.74$ when $W_0 = 12$.

Fits to the surface brightness of globular clusters orbiting our galaxy reveal they have a median concentration index $c = 1.5$, with an interquartile range $c = 1.0$–1.8; in terms of the central potential, the median is $W_0 \approx 7$ and the interquartile range is $W_0 \approx 5$–8. For comparison purposes, the dashed line in Figure 8.1(a) is a Plummer model (Equation 6.107); its core radius is chosen to be $r_c = (5/3)^{1/2} r_K$, in order to match the central curvature of the King models. The Plummer model, by construction, lacks an outer truncation radius r_t.

By construction, the velocity dispersion of a King model is isotropic, with $\langle v_r^2 \rangle = \langle v_\theta^2 \rangle = \langle v_\phi^2 \rangle$. However, for any finite value of W_0, the one-dimensional velocity dispersion only equals the parameter σ at the center of the King model. At larger radii, as shown in Figure 8.1(b), the dispersion decreases, going to zero at the truncation radius. It is only in the limit $W_0 \rightarrow \infty$, or $c \rightarrow \infty$, that the stellar system becomes an isothermal sphere (albeit one with a central core, in this case).

Figure 8.2 shows the projected surface density (in stars per square arcminute) of two relatively nearby globular clusters orbiting within the halo

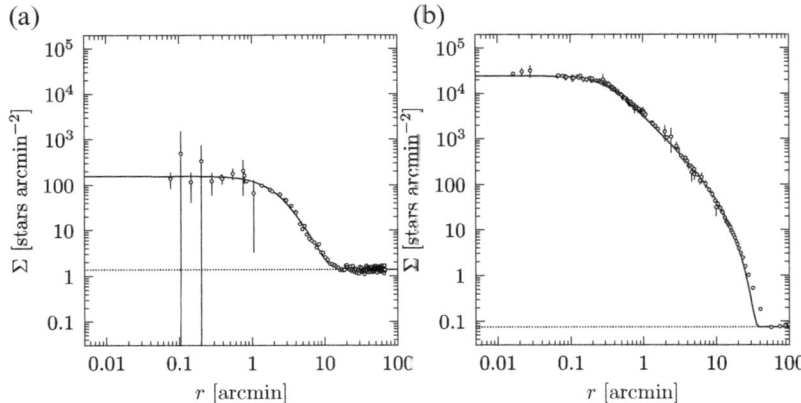

Figure 8.2 Surface density of two nearby globular clusters. (a) NGC 6366 (at d = 3.5 kpc). The curved line is a King model with W_0 = 4.8 and r_t = 20.6 arcmin (21 pc). (b) 47 Tuc (at d = 4.5 kpc). The curved line is a King model with W_0 = 8.6 and r_t = 40.1 arcmin (52 pc). The dotted line in each panel is the estimated background surface density of stars. [Data from de Boer *et al.* 2019]

of the Milky Way Galaxy. These two globular clusters have been chosen to illustrate the typical range of concentration parameters found among globular clusters. The cluster NGC 6366, Figure 8.2(a), is relatively low in concentration, with $W_0 \approx 4.8$, corresponding to $c \approx 1.0$. In contrast, 47 Tucanae is a larger cluster, and is also more concentrated, with $W_0 \approx 8.6$, corresponding to $c \approx 2.0$.

8.1.2 Axisymmetric Systems

Not every stellar system can be adequately approximated as spherical; thus, a distribution function $f(\mathcal{E})$ is too simple to describe all systems. If we consider axisymmetric systems, for instance, we might consider distribution functions of the form $f(\mathcal{E}, j_z)$, where j_z is the component of angular momentum associated with motion around the axis of rotational symmetry. One possible form of the distribution function is

$$f(\mathcal{E}, j_z) = K \left(\frac{j_z}{\sigma R_t} \right)^q \exp \left(\frac{\mathcal{E}}{\sigma^2} \right), \tag{8.30}$$

with the restriction that $f = 0$ when $\mathcal{E} < 0$ or $j_z < 0$. That is, we assume that all stars in the axisymmetric system have the same sign for j_z (which we assume to be positive).[4] The exponent q in Equation 8.30 can be either positive, representing a preference for high angular momentum orbits, or negative, showing a preference for low angular momentum orbits.

---

[4] In other words, this model has no "wrong way" stars orbiting in the opposite sense to the rest.

To simplify the problem, let's assume the axisymmetric stellar system in question is an infinitesimally thin disk, with a four-dimensional phase space $(R, \theta, v_R, v_\theta)$. In these coordinates, we can write $j_z = v_\theta R$, and the distribution function of Equation 8.30 can be written as

$$f(\mathcal{E}, j_z) = K \left(\frac{R}{R_t} \right)^q \exp \left(\frac{\Psi(R)}{\sigma^2} \right) \left(\frac{v_\theta}{\sigma} \right)^q \exp \left(-\frac{v_R^2}{2\sigma^2} - \frac{v_\theta^2}{2\sigma^2} \right). \qquad (8.31)$$

Writing out the distribution function in this way emphasizes that the parameter σ represents the velocity dispersion in the radial direction, which is not necessarily the same as the dispersion in the azimuthal direction.

Integrating Equation 8.31 over the entire (v_R, v_θ) velocity space, we find the surface density of the thin disk,

$$\Sigma(R) = \frac{\sqrt{\pi}}{2} 2^{-q/2} K \sigma^2 \left(\frac{R}{R_t} \right)^q \exp \left(\frac{\Psi(R)}{\sigma^2} \right). \qquad (8.32)$$

Now we have the task of choosing an appropriate relative potential $\Psi(R)$. For a disk with a flat rotation curve ($v_{\text{circ}} = $ constant), the potential must be logarithmic in R, as shown in Equation 7.37. With the appropriate choice of normalization, we can write

$$\Psi(R) = -v_0^2 \ln \left(\frac{R}{R_t} \right), \qquad (8.33)$$

where v_0 is the circular velocity and R_t is the truncation radius of the system. With this logarithmic potential, we find that

$$\exp \left(\frac{\Psi(R)}{\sigma^2} \right) = \left(\frac{R}{R_t} \right)^{-v_0^2/\sigma^2}. \qquad (8.34)$$

The surface density of the thin disk then becomes (substituting into Equation 8.32)

$$\Sigma(R) = \frac{\sqrt{\pi}}{2} 2^{-q/2} K \sigma^2 \left(\frac{R}{R_t} \right)^{q-v_0^2/\sigma^2}. \qquad (8.35)$$

Looking at this surface density, we realize that it represents a Mestel disk, with $\Sigma \propto R^{-1}$, if

$$q = \frac{v_0^2}{\sigma^2} - 1. \qquad (8.36)$$

Since a self-gravitating Mestel disk has a flat rotation curve, choosing this value of q in the distribution function of Equation 8.30 produces an internally consistent, self-gravitating Mestel disk. The limit $q \to \infty$ represents a "cold" disk, with a radial velocity dispersion σ much smaller than the rotation speed v_0; the limit $q \to -1$ produces a "hot" disk, with $\sigma \gg v_0$. Even if a disk has a potential that produces a flat rotation curve, we find there is no requirement that the stars in the disk move tamely along circular orbits; by weighting $f(\mathcal{E}, j_z)$ to lower values of j_z, we can produce a disk of stars on nearly radial orbits.

8.2 Instability and the Jeans Length

A distribution function of the form $f(\vec{r}, \vec{v})$, with no explicit time dependence, implies a system in equilibrium. But is it *stable* equilibrium? The stability of self-gravitating systems was examined quantitatively by James Jeans in 1902, when he published his pioneering paper "The stability of a spherical nebula." Although Jeans used the term "nebula" to mean an interstellar gas cloud, many of his ideas can be applied to a gas in which each particle is a star, rather than an atom or molecule. Although, as Jeans pointed out, rotation can stabilize a gas cloud against gravitational collapse, he looked in particular at the stability of non-rotating spherical systems. We will follow in Jeans' footsteps by considering stellar systems with no net rotation.

The basic equations governing stability against gravitational collapse are the collisionless Boltzmann equation (Equation 6.50),

$$\frac{\partial f_m}{\partial t} + \vec{v} \cdot \vec{\nabla} f_m - \vec{\nabla}\Phi \cdot \frac{\partial f_m}{\partial \vec{v}} = 0, \tag{8.37}$$

and Poisson's equation, which may be written as

$$\nabla^2 \Phi = 4\pi G \int f_m \, d^3 v, \tag{8.38}$$

where $f_m(\vec{r}, \vec{v}, t)$ is the mass distribution function.

Now consider a region within a steady-state stellar system where the mass density ρ is uniform; this region could be within a uniform-density core, or it could simply be a region small enough that the density gradient is imperceptible. Within this region, the steady-state distribution function has the form $f_m = f_0(\vec{v})$ and the gravitational potential has a value $\Phi_0(\vec{r})$ independent of time. Now we slightly perturb the distribution function, leading to

$$f_m(\vec{r}, \vec{v}, t) = f_0(\vec{v}) + f_1(\vec{r}, \vec{v}, t), \tag{8.39}$$

$$\Phi(\vec{r}, t) = \Phi_0(\vec{r}) + \Phi_1(\vec{r}, t). \tag{8.40}$$

The perturbation is assumed to be low in amplitude, with $|f_1| \ll f_0$. We can therefore use linear perturbation theory, inserting the perturbed quantities of Equation 8.40 into the collisionless Boltzmann equation and Poisson's equation and keeping only those terms that are linear in f_1 and Φ_1. This yields

$$\frac{\partial f_1}{\partial t} + \vec{v} \cdot \vec{\nabla} f_1 - \vec{\nabla}\Phi_1 \cdot \frac{\partial f_0}{\partial \vec{v}} = 0, \tag{8.41}$$

$$\nabla^2 \Phi_1 = 4\pi G \int f_1 \, d^3 v. \tag{8.42}$$

For a given unperturbed distribution function $f_0(\vec{v})$, we now want to find which perturbations f_1 have an amplitude that grows with time, representing an instability.

To reduce the computational complexity, we will apply a perturbation that is sinusoidal in space and time, with

$$f_1(\vec{r}, \vec{v}, t) = f_{1,k}(\vec{v})\rho_{1,k} \exp[i(\vec{k} \cdot \vec{r} - \omega t)]. \qquad (8.43)$$

The perturbed potential will have the same wavenumber, frequency, and phase, with

$$\Phi_1(\vec{r}, t) = \Phi_{1,k} \exp[i(\vec{k} \cdot \vec{r} - \omega t)]. \qquad (8.44)$$

Using these sinusoidal perturbations, the collisionless Boltzmann equation (Equation 8.41) gives one relation between f_1 and Φ_1:

$$(\omega - \vec{k} \cdot \vec{v})f_1 = -\vec{k} \cdot \frac{\partial f_0}{\partial \vec{v}} \Phi_1. \qquad (8.45)$$

Poisson's equation (Equation 8.42) gives another relation between f_1 and Φ_1:

$$-k^2 \Phi_1 = 4\pi G \int f_1 \, d^3v. \qquad (8.46)$$

Using Equation 8.45 to substitute for f_1 in Equation 8.46, we find a **dispersion relation** that gives the value of ω for a given $\vec{k}$:

$$-k^2 = 4\pi G \int \frac{\vec{k} \cdot (\partial f_0/\partial \vec{v})}{\vec{k} \cdot \vec{v} - \omega} \, d^3v. \qquad (8.47)$$

In principle, for a given unperturbed distribution function f_0, you could solve this equation for $\omega(\vec{k})$. In practice, it helps to make simplifying assumptions. For example, let's assume a Maxwellian distribution of stellar velocities (Equation 8.14):

$$f_0(\vec{v})d^3v = \frac{\rho_0}{(2\pi\sigma^2)^{3/2}} \exp\left(-\frac{v^2}{2\sigma^2}\right) d^3v, \qquad (8.48)$$

where ρ_0 is the unperturbed, uniform mass density. With this isotropic stellar distribution, the dispersion relation becomes

$$k^2 = \frac{4\pi G\rho_0}{\sigma^2} \int \frac{\vec{k} \cdot \vec{v}}{\vec{k} \cdot \vec{v} - \omega} \exp\left(-\frac{v^2}{2\sigma^2}\right) \frac{d^3v}{(2\pi\sigma^2)^{3/2}}. \qquad (8.49)$$

Let's consider the physics of the problem. A perturbation with $\omega^2 > 0$ will be stable, and have constant amplitude for f_1 and Φ_1. A perturbation with $\omega^2 < 0$ will undergo exponential growth (or perhaps exponential decay) on a timescale $1/|\omega|$. Thus, the wavenumber k that produces $\omega^2 = 0$ is a special case, on the boundary between stability and instability. This special wavenumber is called the **Jeans wavenumber**, as a tribute to James Jeans. By assuming $\omega = 0$

in Equation 8.49, we find that the value of the Jeans wavenumber k_J is given by the relation

$$k_J^2 = \frac{4\pi G\rho_0}{\sigma^2} \int \exp\left(-\frac{v^2}{2\sigma^2}\right) \frac{d^3v}{(2\pi\sigma^2)^{3/2}} = \frac{4\pi G\rho_0}{\sigma^2}. \tag{8.50}$$

This wavenumber corresponds to a **Jeans length**

$$\lambda_J \equiv \frac{2\pi}{k_J} = \left(\frac{\pi\sigma^2}{G\rho_0}\right)^{1/2} = 2.7\,\mathrm{kpc}\left(\frac{\rho_0}{1\,M_\odot\,\mathrm{pc}^{-3}}\right)^{-1/2}\left(\frac{\sigma}{100\,\mathrm{km\,s^{-1}}}\right). \tag{8.51}$$

The Jeans length of a stellar system is similar in form to the Jeans length of an atomic (or molecular) gas. If the gas has sound speed c_s, its Jeans length is

$$\lambda_J = \left(\frac{\pi c_s^2}{G\rho_0}\right)^{1/2}. \tag{8.52}$$

However, there is one striking difference between a gas of stars and a gas of collisional atoms. In the atomic gas, perturbations smaller than the Jeans length are stably propagating sound waves with constant amplitude. In a stellar system, by contrast, perturbations smaller than the Jeans length tend to decrease in amplitude. This damping occurs by a mechanism analogous to Landau damping in collisionless plasmas.

You might wonder whether spherical stellar systems are subject to Jeans instability at all. If the system is in equilibrium, its one-dimensional velocity dispersion is given by the scalar virial theorem as (Equation 6.44)

$$\sigma^2 \sim \frac{GM}{6r_h}, \tag{8.53}$$

where M is the mass of the system and r_h is its half-mass radius. In terms of the mean density $\bar{\rho}$ within the half-mass radius, the virial relation can be written as

$$\sigma^2 \sim \frac{4\pi}{9} G\bar{\rho}r_h^2. \tag{8.54}$$

Using this dispersion and mass density to compute the Jeans length (Equation 8.51), we find that

$$\lambda_J \sim \frac{2\pi}{3}r_h. \tag{8.55}$$

Our naïve assumption would be that, for a spherical system in virial equilibrium, in order for a perturbation to be larger than the Jeans length it would have to be larger than the system itself. A more sophisticated analysis shows that although spherical systems with isotropic velocity dispersions are stable, instabilities exist when the dispersion is nearly radial. Stellar systems can also show instabilities when they are highly flattened along one axis. However, the instabilities of disk galaxies are sufficiently interesting to merit a section all their own.

8.3 Spiral Structure

A striking attribute of disk galaxies is that they display **spiral structure** in their distribution of stars. The discovery of spiral structure in galaxies is usually attributed to William Parsons, the third Earl of Rosse. In 1845, he used his 72-inch aperture telescope (nicknamed the "Leviathan of Parsonstown") to observe the nebula M51 and make the sketch shown in Figure 8.3(a). The distinctive two-armed spiral pattern seen in M51 led to its being called the "Whirlpool Nebula." When Lord Rosse looked at M51, his conclusion was "that such a system should exist, without internal movement, seems to be in the highest degree improbable." When the distinction between gaseous nebulae and stellar systems eventually became clear, astronomers noted that rotationally flattened systems of stars (i.e., disk galaxies) tend to have spiral structure. Sometimes this is a two-armed "grand design" structure like that of M51, seen clearly in Figure 8.3(b); sometimes patchier, multi-armed "flocculent" spirals are seen.[5]

The spiral arms of disk galaxies, when they are deprojected, can be described in terms of their **pitch angle**. Figure 8.4 shows how the pitch angle $\psi_p(R)$ is measured for a perfect spiral curve. A circle of radius R is drawn, centered on the spiral's center. At the point where the circle and spiral curve

(a)   (b)

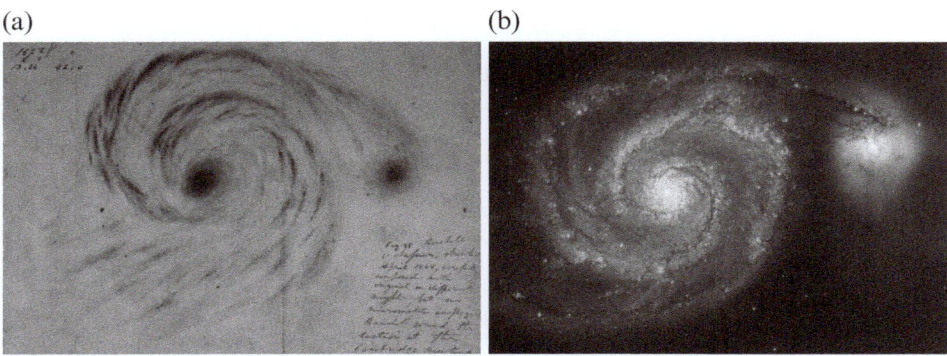

Figure 8.3 M51 is a spiral galaxy at a distance $d \approx 8.6\,\mathrm{Mpc}$. (a) Drawing by Lord Rosse of M51, done in April 1845. [Birr Castle archives] (b) *Hubble Space Telescope* image of M51, taken in January 2005. The image size is 9.6×6.7 arcmin, corresponding to $\sim 24 \times 17\,\mathrm{kpc}$ at the distance of M51. [NASA, ESA, S. Beckwith (STScI), Hubble Heritage Team (STScI/AURA)]

[5] The word "flocculent" comes from the Latin *floccus*, meaning a tuft of wool.

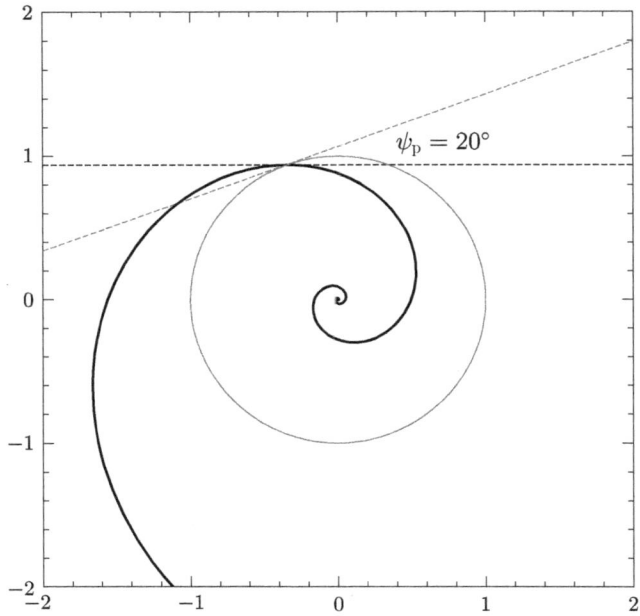

Figure 8.4 Measuring the pitch angle of a spiral. In this case, the constant pitch angle of the logarithmic spiral (dark gray) is $\psi_p = 20°$.

intersect, two lines are drawn: one is tangent to the circle, while the other is tangent to the spiral. The angle ψ_p between the two tangent lines is the pitch angle; by definition, the pitch angle lies in the range $0 < \psi_p < 90°$.

A spiral whose pitch angle is constant with R is called a **logarithmic spiral**. In polar coordinates, the equation for a logarithmic spiral is

$$R(\theta) = a \exp(b\theta) \qquad (8.56)$$

or, equivalently,

$$\theta(R) = \frac{1}{b} \ln\left(\frac{R}{a}\right). \qquad (8.57)$$

The pitch angle for a logarithmic spiral is given by the relation

$$\tan \psi_p = \frac{1}{R}\left|\frac{d\theta}{dR}\right|^{-1} = |b|. \qquad (8.58)$$

Most spiral galaxies do not have perfect logarithmic spirals; only ~40% of spiral galaxies have ψ_p constant to within 20% over the range of R for which the spiral arms can be traced. However, if the pitch angle is averaged over the entire range of R, it is found that spiral galaxies with large bulges (type Sa) have a median pitch angle $\psi_p \sim 10°$. By contrast, small-bulged spiral galaxies (type Sd) are more loosely wound, with a median pitch angle as high as $\psi_p \sim 30°$. The Milky Way Galaxy, which is type Sb, has an average pitch angle $\psi_p \approx 13°$. Disk galaxies with little or no gas (type S0) do not show spiral structure; a full treatment of spiral arms must necessarily include the effect of gas dynamics. However, let's see what we can deduce about spiral structure just by using the tools of stellar dynamics.

Spiral arms cannot contain the same stars for the entire lifetime of a disk galaxy (which for the Milky Way Galaxy, for instance, is $t_{sys} \sim 10\,\mathrm{Gyr}$). Instead, long-lived spiral arms must be **density waves**. To see why, take all the stars that lie close to a radial line $\theta = \theta_0$ at time $t = 0$ and tag them. If these stars are on nearly circular orbits, we can compute what the angle $\theta(R, t)$ of the tagged stars must be at some later time. If we define θ as increasing in the direction of orbital motion then, at $t > 0$,

$$\theta(R, t) = \theta_0 + \Omega(R)t, \tag{8.59}$$

where $\Omega(R)$ is the angular speed of a star on an orbit of radius R. For a galaxy with a flat rotation curve, $\Omega(R) = v_{circ}/R$; this leads to a pitch angle $\psi_p(R, t)$ given by the relation

$$\tan \psi_p = \frac{R}{v_{circ}t} = 0.0341 \left(\frac{v_{circ}}{235\,\mathrm{km\,s^{-1}}} \right)^{-1} \left(\frac{R}{8.2\,\mathrm{kpc}} \right) \left(\frac{t}{1\,\mathrm{Gyr}} \right)^{-1}. \tag{8.60}$$

To have a pitch angle $\psi_p \approx 13°$, comparable to that of the Milky Way Galaxy, the age of a spiral arm would have to be

$$t = 148\,\mathrm{Myr} \left(\frac{v_{circ}}{235\,\mathrm{km\,s^{-1}}} \right)^{-1} \left(\frac{R}{8.2\,\mathrm{kpc}} \right) \left(\frac{\tan \psi_p}{\tan 13°} \right)^{-1}. \tag{8.61}$$

Some spiral structures, such as the short arms typical of flocculent spirals, may be relatively short-lived stellar associations sheared into a spiral shape. If the angular speed Ω decreases with R, then spiral arms of this kind must be **trailing** arms. Using the usual convention that θ increases in the direction of orbital motion, then a trailing arm is one with $d\theta/dR < 0$. Conversely, a **leading** spiral arm has $d\theta/dR > 0$.

Although flocculent spirals might have short-lived arms consisting of a fixed population of stars, grand design spirals like those of M51 are better explained as being density waves. The concept of spiral arms as density waves was proposed by Frank Shu and C. C. Lin in the 1960s, and thus is sometimes called the Lin–Shu density wave theory. The density wave theory states that spiral arms are regions of a galaxy's disk where the surface density $\Sigma(R, \theta)$ is greater than the azimuthally averaged density $\Sigma_0(R)$. However, this overdensity is not provided by a fixed set of stars. Instead, the density wave rotates around the disk center with a nearly uniform angular speed Ω_p; the uniformity of this **pattern speed** eliminates the winding problem for spiral arms. Any individual star in the disk orbits at a speed $\Omega(R)$. If $\Omega > \Omega_p$, the star will overtake the spiral arm; if $\Omega < \Omega_p$, the arm will overtake the star. In either case, the star spends a limited time within the spiral arm.

If a spiral density wave moves with a constant pattern speed Ω_p, then there will be resonances at certain special radii within the disk. For instance, there may be a **corotation radius** R_{co}, where

$$\Omega(R_{co}) = \Omega_p. \tag{8.62}$$

A star on a circular orbit with this radius has the ability to "surf" on the density wave. Other special radii, for stars on almost circular orbits, are connected with the epicycle frequency (Equation 7.63)

$$\kappa(R) = 2\Omega(R)\left[1 + \frac{1}{2}\frac{d\ln\Omega}{d\ln R}\right]^{1/2}. \tag{8.63}$$

If a spiral density wave has m arms, then a star at $R \neq R_{co}$ will encounter the density wave with a frequency

$$\omega_{enc} = m|\Omega_p - \Omega(R)|. \tag{8.64}$$

The **inner Lindblad resonance** is located at the radius where

$$m[\Omega(R_{il}) - \Omega_p] = \kappa(R_{il}). \tag{8.65}$$

This tells us that at the inner Lindblad resonance, every time a star encounters the density wave, it is at the same point on its epicycle, thus creating a resonance. There is also an outer Lindblad resonance, where

$$m[\Omega_p - \Omega(R_{ol})] = \kappa(R_{ol}). \tag{8.66}$$

Again, at the outer Lindblad resonance, every time a star encounters the density wave, it is at the same point on its epicycle.

Unfortunately for astronomers who want to determine the location of the corotation and Lindblad resonances, their effects on the stellar distribution in the disk can be subtle. However, it is generally true that outside the outer Lindblad resonance, where Ω_p becomes large compared to both $\Omega(R)$ and $\kappa(R)$, density waves tend to fade away. In addition, the Lindblad resonances tend to be "hilda-like" resonances, associated with an excess rather than a deficit of stars.

In the Milky Way Galaxy, the pattern speed of the spiral arms is estimated to be $\Omega_p \approx 25\,\text{km s}^{-1}\,\text{kpc}^{-1} \approx 8.1 \times 10^{-16}\,\text{s}^{-1}$. Assuming a flat rotation curve, this places our galaxy's corotation radius at

$$R_{co} = 9.4\,\text{kpc}\left(\frac{v_{circ}}{235\,\text{km s}^{-1}}\right)\left(\frac{\Omega_p}{25\,\text{km s}^{-1}\,\text{kpc}}\right)^{-1}. \tag{8.67}$$

Since $\kappa = \sqrt{2}\Omega$ for a galaxy with a flat rotation curve, this places the inner and outer Lindblad resonance for a two-armed spiral (Equations 8.65 and 8.66) at

$$R_{\rm il} = \left(1 - \frac{1}{\sqrt{2}}\right) \frac{v_{\rm circ}}{\Omega_{\rm p}} \approx 0.29\, R_{\rm co}, \qquad (8.68)$$

$$R_{\rm ol} = \left(1 + \frac{1}{\sqrt{2}}\right) \frac{v_{\rm circ}}{\Omega_{\rm p}} \approx 1.71\, R_{\rm co}. \qquad (8.69)$$

For a corotation radius $R_{\rm co} = 9.4\,{\rm kpc}$, these correspond to $R_{\rm il} \approx 2.8\,{\rm kpc}$ and $R_{\rm ol} \approx 16\,{\rm kpc}$.

8.4 Stability of Density Waves

A sound wave traveling through a gas can be treated as a low-amplitude pressure perturbation, traveling with a characteristic sound speed $c_{\rm s}$. Similarly, a density wave traveling through a thin stellar disk can be treated as a low-amplitude perturbation to the surface density Σ of the disk. For a quantitative study of density waves, let's start with an idealized disk galaxy. Our ideal galaxy is assumed to be an axisymmetric, steady-state, thin disk, with surface mass density $\Sigma_0(R)$ and a gravitational potential $\Phi_0(R)$ in the midplane of the disk. All the stars in the disk are assumed to orbit in the same sense about the disk center. If a star is on a perfectly circular orbit in the midplane, its angular speed is (Equation 7.42)

$$\Omega_0(R) = \left[\frac{1}{R}\frac{d\Phi_0}{dR}\right]^{1/2}. \qquad (8.70)$$

If a star is on a slightly non-circular orbit, its epicyclic motion has the frequency (Equation 7.63)

$$\kappa_0(R) = 2\Omega_0(R)\left[1 + \frac{1}{2}\frac{d\ln\Omega_0}{d\ln R}\right]^{1/2}. \qquad (8.71)$$

To the axisymmetric thin disk, we add a surface density perturbation that can be written generally as

$$\Sigma(R,\theta,t) = \Sigma_0(R) + \Sigma_1(R,\theta,t), \qquad (8.72)$$

where $|\Sigma_1| \ll \Sigma_0$ everywhere in the disk. The bulk velocity $\vec{u}$ of the stars in the disk will now be slightly perturbed from the bulk velocity $u_{\theta 0}$ in the axisymmetric, steady-state disk:

$$\vec{u}(R,\theta,t) = u_{\theta 0}(R)\hat{\theta} + [u_{R1}(R,\theta,t)\hat{R} + u_{\theta 1}(R,\theta,t)\hat{\theta}]. \qquad (8.73)$$

The potential in the midplane of the disk will also be slightly perturbed, with

$$\Phi(R, \theta, t) = \Phi_0(R) + \Phi_1(R, \theta, t). \tag{8.74}$$

When we take our unperturbed axisymmetric disk and stamp it with some density perturbation Σ_1, we can then ask whether the perturbation will grow in amplitude, decay in amplitude, or stably propagate as a nearly constant-amplitude density wave. The stability analysis starts with the mass continuity equation for a thin disk; this can be written in polar coordinates as

$$\frac{\partial \Sigma}{\partial t} + \frac{1}{R}\frac{\partial}{\partial R}(R\Sigma u_R) + \frac{1}{R}\frac{\partial}{\partial \theta}(\Sigma u_\theta) = 0. \tag{8.75}$$

Substituting in the values for the perturbed surface density and bulk velocity (Equations 8.72 and 8.73), then taking only those terms linear in Σ_1, u_{R1}, and $u_{\theta 1}$, the linearized mass continuity equation becomes

$$\frac{\partial \Sigma_1}{\partial t} + \frac{1}{R}\frac{\partial}{\partial R}(R\Sigma_0 u_{R1}) + \Omega_0\frac{\partial \Sigma_1}{\partial \theta} + \frac{\Sigma_0}{R}\frac{\partial u_{\theta 1}}{\partial \theta} = 0. \tag{8.76}$$

The *momentum* continuity equations for a disk, as given in Equations 6.67–6.69, are complicated by the fact that the velocity dispersion can be aniso-tropic. To study instabilities in as simple a system as possible, let's assume a disk made of gas in which the sound speed c_s represents an isotropic veloc-ity dispersion. For a gaseous disk, the momentum continuity equation in the radial direction becomes (in its linearized form)

$$\frac{\partial u_{R1}}{\partial t} + \Omega_0\frac{\partial u_{R1}}{\partial \theta} - 2\Omega_0 u_{\theta 1} = -\frac{\partial \Phi_1}{\partial R} - \frac{c_s^2}{\Sigma_0}\frac{\partial \Sigma_1}{\partial R}. \tag{8.77}$$

The two terms on the right-hand side of Equation 8.77 represent the per-turbed gravitational force and the perturbed pressure force. The momentum continuity equation in the azimuthal direction similarly becomes

$$\frac{\partial u_{\theta 1}}{\partial t} + \frac{\kappa_0^2}{2\Omega_0}u_{R1} + \Omega_0\frac{\partial u_{\theta 1}}{\partial \theta} = -\frac{1}{R}\frac{\partial \Phi_1}{\partial \theta} - \frac{c_s^2}{\Sigma_0 R}\frac{\partial \Sigma_1}{\partial \theta}. \tag{8.78}$$

Finally, Poisson's equation for perturbations to an infinitesimally thin disk becomes

$$\nabla^2\Phi_1 = 4\pi G\Sigma_1\delta(z), \tag{8.79}$$

where Σ_1 represents a mass surface density and $\delta(x)$ is the Dirac delta distribution.

If we like, we can stamp a rotating disk with whatever (low-amplitude) density perturbation takes our fancy. However, since real disk galaxies char-acteristically have spiral density perturbations. we want a functional form for Σ_1 that mimics a spiral density wave. An appropriate form for a spiral with m arms is

$$\Sigma_1(R, \theta, t) = S(R)\exp[if(R)]\exp[i(m\theta - \omega t)], \tag{8.80}$$

where we adopt the usual convention that the azimuthal coordinate θ increases in the direction of the stars' unperturbed orbital motion. The function $S(R)$ is the azimuthally averaged density profile of the density perturbation; it captures the gradual change in density as you move outward along a spiral arm. The **shape function** $f(R)$ represents the rapid inter-arm variation as you move outward along a purely radial path. The radial wavenumber k associated with the spiral arms is $k(R) \equiv df/dR$. At a fixed time t, Equation 8.80 tells us that $f(R) + m\theta = $ constant as we move outward along the crest of a spiral density wave. This implies that at any point on the wave crest,

$$k = \frac{df}{dR} = -m\frac{d\theta}{dR}. \tag{8.81}$$

Thus, a trailing spiral arm, with $d\theta/dR < 0$, has $k > 0$; conversely, a leading spiral arm has $k < 0$.

For a logarithmic spiral, as shown in Figure 8.4, the relation between the pitch angle ψ_p and radial wavenumber k is

$$\tan \psi_p = \frac{1}{R}\left|\frac{d\theta}{dR}\right|^{-1} = \frac{m}{|k|R}. \tag{8.82}$$

Notice that at a fixed radius R, Equation 8.80 tells us that the density wave rotates in such a way that a point on the wave crest has $m\theta - \omega t = $ constant. Thus, the pattern speed of the density wave is $\Omega_p = d\theta/dt = \omega/m$.

Since we have linearized the continuity equations, and since Poisson's equation is intrinsically linear, if we impose a density perturbation Σ_1 that is sinusoidal in θ and t, we will produce velocity perturbations u_{R1} and $u_{\theta 1}$ and a potential perturbation Φ_1 that are also sinusoidal, with the same values of m and ω. This means that we can write $\partial F_1/\partial\theta = imF_1$ and $\partial F_1/\partial t = -i\omega F_1$, where the function F can be Σ, u_R, u_θ, or Φ. We now simplify the equations still further by adopting the WKB approximation. The WKB approximation was adopted by Wentzel, Kramers, and Brillouin in the context of semi-classical quantum mechanics. In the context of spiral structure, however, it amounts to the assumption that the spiral arms are tightly wound, with $\tan\psi_p \ll 1$; for a roughly logarithmic spiral, this is equivalent to assuming $|k| \gg m/R$ (from Equation 8.82). In the WKB approximation, the radial derivatives are dominated by the high-k fluctuations going from crest to trough to crest. Thus, it is safe to write $\partial F_1/\partial R \approx ikF_1$.

Given the WKB approximation, the mass continuity equation for a tightly wound spiral becomes

$$(m\Omega_0 - \omega)\Sigma_1 + k\Sigma_0 u_{R1} = 0. \tag{8.83}$$

The momentum continuity equations become

$$(m\Omega_0 - \omega)u_{R1} + i(2\Omega_0)u_{\theta 1} = -k\left(\Phi_1 + \frac{c_s^2}{\Sigma_0}\Sigma_1\right), \tag{8.84}$$

$$(m\Omega_0 - \omega)u_{\theta 1} - i\left(\frac{\kappa_0^2}{2\Omega_0}\right)u_{R1} = 0. \tag{8.85}$$

The recurring term $m\Omega_0 - \omega = m(\Omega_0 - \Omega_p)$ in the continuity equations represents the frequency with which a star on a circular orbit encounters a spiral arm. Finally, Poisson's equation in the disk plane ($z = 0$) becomes

$$\Phi_1 = -\frac{2\pi G}{|k|}\Sigma_1. \tag{8.86}$$

Equations 8.83–8.86 are four equations in the four unknowns $\Sigma_1, u_{R1}, u_{\theta 1}$, and Φ_1. With simple algebra, you can confirm there is a self-consistent solution for these quantities only if the frequency ω obeys the dispersion relation

$$(m\Omega_0 - \omega)^2 = \kappa_0^2 - 2\pi G\Sigma_0|k| + c_s^2 k^2. \tag{8.87}$$

We can now take an initial unperturbed disk with known values of $\Sigma_0(R)$, $c_s(R)$, and $\Omega_0(R)$.[6] Then we add a density wave in the form of an m-armed spiral with wavenumber $k(R)$. The dispersion relation of Equation 8.87 then tells us what range of R (if any) produces a real value of ω, corresponding to stably propagating density waves.

The dispersion relation of Equation 8.87 can be written slightly more elegantly as

$$s^2 = 1 - 2\frac{|k|}{k_{crit}} + Q^2\frac{k^2}{k_{crit}^2}, \tag{8.88}$$

where the dimensionless frequency is

$$s \equiv \frac{m\Omega_0 - \omega}{\kappa_0}, \tag{8.89}$$

which has the value $s = 0$ at corotation and $s = \pm 1$ at the inner and outer Lindblad resonances. Since m, Ω_0, and κ_0 are all real, then if s is a real number, $\omega(k)$ must also be real, and the density wave of wavenumber k propagates stably. In Equation 8.88, the critical wavenumber k_{crit} is

$$k_{crit} \equiv \frac{\kappa_0^2}{\pi G\Sigma_0} \tag{8.90}$$

$$= 6.40 \times 10^{-22}\,\text{cm}^{-1}\left(\frac{\kappa_0}{40\,\text{km s}^{-1}\,\text{kpc}^{-1}}\right)^2\left(\frac{\Sigma_0}{60\,M_\odot\,\text{pc}^{-2}}\right)^{-1},$$

[6] Knowing $\Omega_0(R)$ lets you compute $\kappa_0(R)$ using Equation 7.63.

scaling to properties of the Milky Way disk in the solar neighborhood. The equivalent critical wavelength is

$$\lambda_{\mathrm{crit}} = \frac{2\pi}{k_{\mathrm{crit}}} = 3.18\,\mathrm{kpc} \left(\frac{\kappa_0}{40\,\mathrm{km\,s^{-1}\,kpc^{-1}}}\right)^{-2} \left(\frac{\Sigma_0}{60\,\mathrm{M_\odot\,pc^{-2}}}\right). \tag{8.91}$$

Finally, the dimensionless parameter Q in Equation 8.88 is **Toomre's Q parameter**, defined as

$$Q \equiv \frac{c_s \kappa_0}{\pi G \Sigma_0} \tag{8.92}$$

$$= 0.49 \left(\frac{c_s}{10\,\mathrm{km\,s^{-1}}}\right) \left(\frac{\kappa_0}{40\,\mathrm{km\,s^{-1}\,kpc^{-1}}}\right) \left(\frac{\Sigma_0}{60\,\mathrm{M_\odot\,pc^{-2}}}\right)^{-1}.$$

The importance of Toomre's Q parameter becomes more apparent when we look at the dispersion relation (Equation 8.88) for the special case $Q = 1$. In this case,

$$s^2 = \left(1 - \frac{|k|}{k_{\mathrm{crit}}}\right)^2, \tag{8.93}$$

and $s^2 > 0$ for all wavenumbers except $|k| = k_{\mathrm{crit}}$, which has $s = 0$. Thus, $Q = 1$ represents a disk that is just barely stable.

When $Q > 1$, the disk is a "hot" disk, with every wavenumber k corresponding to $s^2 > 0$; in a hot disk, the pressure associated with the high sound speed stabilizes the system, just as in a system stabilized against Jeans collapse. When $Q < 1$, the resulting "cool" disk is unstable for some values of the wavenumber k. Solving the dispersion relation (Equation 8.88) in the limit $Q^2 \ll 1$, we find that $s^2 < 0$, representing unstable density waves, over the range of wavenumbers

$$\frac{1}{2}\left[1 + \frac{Q^2}{4}\right] < \frac{|k|}{k_{\mathrm{crit}}} < \frac{2}{Q^2}. \tag{8.94}$$

In this range, the minimum value of s^2, corresponding to the fastest-growing unstable wave, occurs at a wavenumber

$$k_{\mathrm{unst}} = \frac{k_{\mathrm{crit}}}{Q^2} = \frac{\pi G \Sigma_0}{c_s^2}, \tag{8.95}$$

corresponding to a wavelength

$$\lambda_{\mathrm{unst}} = \frac{2\pi}{k_{\mathrm{unst}}} = 0.78\,\mathrm{kpc} \left(\frac{c_s}{10\,\mathrm{km\,s^{-1}}}\right)^2 \left(\frac{\Sigma_0}{60\,\mathrm{M_\odot\,pc^{-2}}}\right)^{-1}. \tag{8.96}$$

Note that even in the limit of an "absolute zero" disk, with $Q = 0$, density waves with $|k| < k_{\mathrm{crit}}/2$, corresponding to $\lambda > 2\lambda_{\mathrm{crit}}$, will be stabilized by the disk's differential rotation. Figure 8.5 shows the quadratic dispersion relation

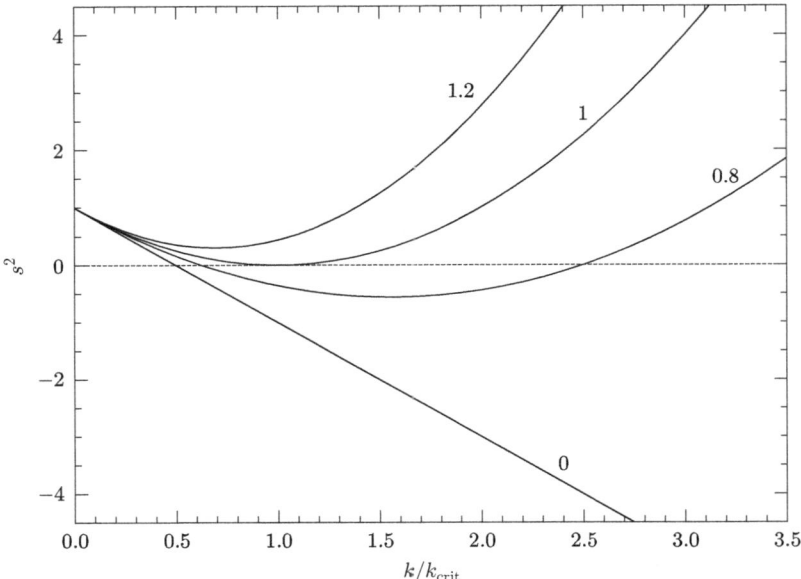

Figure 8.5 Dispersion relation of Equation 8.88 for a warm disk ($Q = 1.2$), a critical disk ($Q = 1$), a cool disk ($Q = 0.8$), and an "absolute zero" disk ($Q = 0$). Waves are unstable below the dashed horizontal line.

of Equation 8.88 for different values of Q, plus the linear dispersion relation for $Q = 0$.

For a stellar disk, the analysis is more complicated than for a gas disk. In a stellar disk, the stabilization is provided by the dispersion σ_{RR} in the radial direction. Toomre's Q parameter for a stellar disk turns out to be

$$Q_\star = \frac{\sigma_{RR}\kappa_0}{3.36G\Sigma_0}. \tag{8.97}$$

The azimuthal dispersion $\sigma_{\theta\theta}$ is irrelevant for the stability of a density wave; it is the radial dispersion σ_{RR} that determines whether a disk is hot and stable ($Q_\star > 1$) or cold and unstable ($Q_\star < 1$).

If the rotation of spiral density waves has made you a bit dizzy, it is also possible to apply the Q parameter to axisymmetric perturbations. By setting $m = 0$ in Equation 8.80, we produce a perturbation of the form

$$\Sigma_1(R, t) = H(R)\exp[if(R)]\exp(-i\omega t) \tag{8.98}$$

with radial wavenumber $k(R) = df/dR$. In the case of axisymmetric perturbations, the dispersion relation of Equation 8.88 reduces to

$$\frac{\omega^2}{\kappa_0^2} = 1 - 2\frac{|k|}{k_{\text{crit}}} + Q^2\frac{k^2}{k_{\text{crit}}^2}, \tag{8.99}$$

where k_{crit} and Q have the same value as they do for spiral perturbations.

Exercises

8.1   The Jeans length can be calculated for any object of known mass density
and sound speed (or velocity dispersion).
(a) What is the Jeans length in the Sun's photosphere? What is the Jeans
length at the Sun's center? Give your answers in units of the Sun's
radius $R_\odot$. [Bonus questions: What source did you use for ρ and T
inside the Sun? Why do you regard this as a reliable source?]
(b) What is the Jeans length of the Earth's atmosphere at sea level at
an average air temperature? How does this compare to the atmosphere's scale height? [You may look up the scale height rather than
calculating it.]
(c) What was the Jeans length of Sir James Jeans?
8.2   Show that the dispersion relation of Equation 8.49 yields a frequency

$$\omega^2 \approx -4\pi G \rho_0 \tag{8.100}$$

in the limit $k \rightarrow 0$ and the wavelength of the perturbation becomes
arbitrarily large.

9

Encounters Between Stellar Systems

Ships that pass in the night, and speak each other in passing,
Only a signal shown and a distant voice in the darkness;
So on the ocean of life we pass and speak one another,
Only a look and a voice, then darkness again and a silence.
                                Henry Wadsworth Longfellow (1807–1882)
                                "The Theologian's Tale: Elizabeth" [1873]

Galaxies don't live in isolation; mergers between galaxies can occur. The Milky Way Galaxy and the Andromeda Galaxy, for instance, are falling toward each other on an orbit that is consistent with being nearly radial. The relative speed of the two galaxies is $v_r \approx 109 \, \text{km s}^{-1} \approx 112 \, \text{kpc Gyr}^{-1}$, while their current separation (center-to-center) is $d \approx 770 \, \text{kpc}$. Disregarding the gravitational attraction between them, the time elapsed until their merger would be $t \sim d/v_r \sim 7 \, \text{Gyr}$. Their mutual gravitational attraction will hasten their merger; simulations indicate it will happen $t \sim 4.5 \, \text{Gyr}$ from now.[1] In crowded locations like the centers of rich galaxy clusters, mergers are relatively common, building up the brightest cluster galaxies (BCGs) seen at the center of clusters.

A prominent example of two spiral galaxies undergoing a merger is the pair of galaxies known as the Antennae (also known as NGC 4038 and NGC 4039). Figure 9.1 shows the elongated, curved features that give the Antennae their name. These features are tidal tails, containing stars that have been stripped from their parent galaxy by differential tidal forces. The central regions of the two galaxies also show prominent star formation. (There goes the assumption that stars are not created or destroyed!) During a galaxy merger, physical impacts between stars are rare; however, impacts between

---

[1] By coincidence, this is also the approximate time that will elapse until the Sun becomes a red giant.

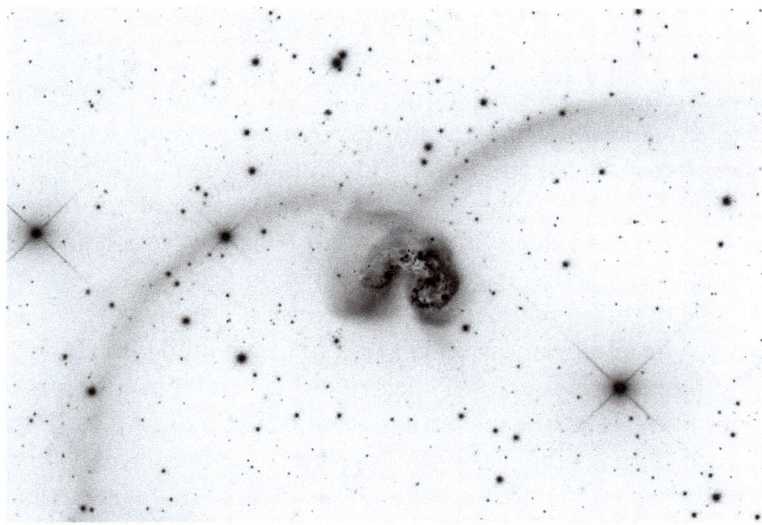

Figure 9.1 Image of the Antennae (NGC 4038 / NGC 4039), a pair of interacting galaxies at $d \sim 22\,\mathrm{Mpc}$. The image size is $18.7 \times 13.1\,\mathrm{arcmin}$, corresponding to $\sim 120 \times 84\,\mathrm{kpc}$ at the distance of the Antennae. [Robert Gendler]

interstellar gas clouds, with their much larger cross-sections, can trigger highly visible outbursts of star formation.

To avoid the complications provided by star formation, let's consider the merger of two gas-free stellar systems. If two point masses move along hyperbolic orbits ($\epsilon > 0$) relative to each other, then once they pass their periapsis, they will continue to move apart to infinity. However, if two galaxies have an impact parameter b that is small compared to the size of the galaxies, they will overlap at the time of periapsis. During the epoch of closest approach, energy will be converted from the orbital kinetic energy (associated with the relative motion of the two galaxies) into internal kinetic energy (associated with the motion of stars within each galaxy). If enough energy is converted, then the two galaxies will be gravitationally bound to each other. Since a close encounter between galaxies converts ordered motion into random motion, it increases the entropy of the system and can be thought of as a frictional process.

Frequently, the merger of two galaxies is a "train wreck," best modeled numerically. However, there are some limiting cases where the situation is simpler. A dense compact system moving through a larger, lower-density stellar system can be studied using the dynamical friction approximation. High-speed encounters between stellar systems can be studied using the impulse approximation. Relatively distant encounters can be studied using the purely tidal approximation.

9.1 Dynamical Friction

In the 1940s, Subrahmanyan Chandrasekhar did a comprehensive mathematical study of **dynamical friction**, the process by which a compact body passing through a background distribution of stars loses its kinetic energy. The energy lost by the compact object goes to "heat up" the background of stars, in the sense of increasing its velocity dispersion. To treat the problem of dynamical friction mathematically, start with a compact object of mass M. The term "compact object" is a broadly inclusive one: the object may be a black hole, but it may also be an ultramassive star or a tightly bound globular cluster. In a frame of reference where the bulk velocity of the background stars is $\vec{u} = 0$, the compact mass M is moving with a velocity $\vec{v}_M$, as shown in Figure 9.2. In the absence of the compact object's perturbing influence, the stars have a number density $n_\star$ and a corresponding mass density $\rho_\star = m_\star n_\star$, where $m_\star$ is the average stellar mass. We will assume that $M \gg m_\star$; in this limit, as we will see, the effects of dynamical friction depend on the stellar mass density $\rho_\star$, and not on the mass of the individual stars; thus, we may adopt the useful fiction that all stars have an identical mass $m_\star$.

Start by considering the interaction of the compact mass M with a single star of mass $m_\star \ll M$. Initially, when the two objects are widely separated, their relative speed is v and their impact parameter is b. The angle $\Delta\theta$ through which the two bodies are deflected by their encounter is given by Equation 1.42:

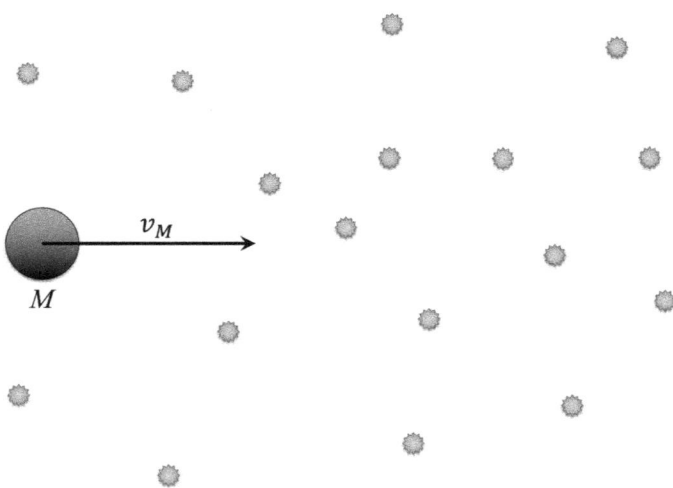

Figure 9.2 A compact object of mass M moves with speed v_M relative to a background of lower-mass stars with number density $n_\star$ and velocity dispersion σ.

$$\sin\left(\frac{\Delta\theta}{2}\right) = \frac{1}{e}. \tag{9.1}$$

For this hyperbolic encounter, the eccentricity e is, from Equation 6.8,

$$e = (1 + b^2/b_{cge}^2)^{1/2}. \tag{9.2}$$

The critical impact parameter for a 90° deflection is, from Equation 6.9,

$$b_{cge} \equiv \frac{G(M + m_\star)}{v^2} \approx 90\,\text{au}\left(\frac{M}{10^3\,\text{M}_\odot}\right)\left(\frac{v}{100\,\text{km s}^{-1}}\right)^{-2}. \tag{9.3}$$

We can combine Equations 9.1 and 9.2 to write

$$\sin\left(\frac{\Delta\theta}{2}\right) = \frac{1}{(1 + b^2/b_{cge}^2)^{1/2}}. \tag{9.4}$$

It then follows that

$$\cos\left(\frac{\Delta\theta}{2}\right) = \frac{b/b_{cge}}{(1 + b^2/b_{cge}^2)^{1/2}}. \tag{9.5}$$

Finally, we can use a standard trigonometric identity to write

$$\sin(\Delta\theta) = 2\sin\left(\frac{\Delta\theta}{2}\right)\cos\left(\frac{\Delta\theta}{2}\right) = 2\frac{b/b_{cge}}{1 + b^2/b_{cge}^2}. \tag{9.6}$$

Thus, even a distant gravitational encounter, with $b \gg b_{cge}$, will produce a small deflection, with $\Delta\theta \approx 2b_{cge}/b$ in this limit.

As a result of the encounter with a single star, the compact object will receive a velocity kick in the direction perpendicular to its initial direction of motion. The magnitude of this velocity change is

$$\Delta v_{M\perp} = \frac{m_\star}{M + m_\star}v\sin(\Delta\theta) \approx \frac{2m_\star v}{M}\frac{b/b_{cge}}{1 + b^2/b_{cge}^2}. \tag{9.7}$$

This perpendicular change in velocity is maximized when $b = b_{cge}$. Since the compact object and star acquire a non-zero relative velocity in the perpendicular direction, to conserve energy they must decrease their relative velocity in the parallel direction. Parallel to its initial direction of motion, the compact mass has a velocity loss

$$\Delta v_{M\parallel} \approx -\frac{2m_\star v}{M}\frac{1}{1 + b^2/b_{cge}^2}. \tag{9.8}$$

This parallel loss of velocity has its magnitude maximized in the limit $b \to 0$.

Now let's send our compact mass M through a population of stars with uniform density $n_\star$ and steady-state distribution function $f(\vec{r}_m, \vec{v}_m, t) = n_\star f_v(\vec{v}_m)$. In the frame of reference where the bulk velocity of the stars is zero, the velocity of the compact object is $\vec{v}_M$. The speed of the compact object relative to any individual star is then $v = |\vec{v}_M - \vec{v}_m|$, where $\vec{v}_m$ is the star's velocity. The rate at which the compact object encounters stars with an impact parameter in the range $b \to b + db$ and a stellar velocity in the element $d^3 v_m$ centered on a velocity $\vec{v}_m$ is

$$\frac{dN_{\mathrm{enc}}}{dt} = |\vec{v}_M - \vec{v}_m| n_\star \, f_v(\vec{v}_m) \, d^3 v_m (2\pi b \, db). \tag{9.9}$$

On average, the velocity changes $\Delta v_{M\perp}$ from all stellar encounters will cancel out; given the uniform density $n_\star$, for every kick to the right, there will be a kick to the left. However, the changes $\Delta v_{M\|}$ will add together, since they all act to slow the compact object. Thus, the rate of slowing will be

$$\frac{dv_M}{dt} = \frac{dN_{\mathrm{enc}}}{dt} \cdot \Delta v_{M\|} \tag{9.10}$$

$$\approx -4\pi n_\star \frac{m_\star}{M} |\vec{v}_M - \vec{v}_m|^2 \, f_v(\vec{v}_m) \, d^3 v_m \cdot \frac{b \, db}{1 + b^2/b_{\mathrm{cge}}^2}, \tag{9.11}$$

making use of Equation 9.8. If we now integrate over all impact parameters, we must acknowledge that real stellar systems have a finite size, and that there exists a maximum possible impact parameter b_{max} comparable to the radius of the stellar system in question. With the reasonable assumption that $b_{\mathrm{max}} \gg b_{\mathrm{cge}}$, we find that

$$\int_0^{b_{\mathrm{max}}} \frac{b \, db}{1 + b^2/b_{\mathrm{cge}}^2} \approx b_{\mathrm{cge}}^2 \ln \Lambda, \tag{9.12}$$

where[2] $\Lambda \equiv b_{\mathrm{max}}/b_{\mathrm{cge}}$. Substituting this result into Equation 9.11, we find that the deceleration of the compact object, integrated over all possible impact parameters, is

$$\frac{dv_M}{dt} \approx -\frac{4\pi \rho_\star}{M} |\vec{v}_M - \vec{v}_m|^2 \, f_v(\vec{v}_m) \, d^3 v_m \cdot b_{\mathrm{cge}}^2 \ln \Lambda, \tag{9.13}$$

where $\rho_\star = m_\star n_\star$. Using the definition of the critical impact parameter b_{cge}, Equation 9.13 becomes

$$\frac{dv_M}{dt} = -4\pi G^2 \rho_\star M \ln \Lambda \frac{f_v(\vec{v}_m) \, d^3 v_m}{|\vec{v}_M - \vec{v}_m|^2}. \tag{9.14}$$

To find the net deceleration of the compact object, we need to integrate Equation 9.14 over the entire stellar velocity distribution f_v. In general, the

---

[2] The term $\ln \Lambda$ is commonly known as the Coulomb logarithm; it was first introduced in the context of Coulomb interactions between charged particles in a plasma.

parameter $\Lambda = b_{\mathrm{max}}/b_{\mathrm{cge}}$ depends on a star's velocity $\vec{v}_m$. Making use of Equation 9.3, we find

$$\ln \Lambda \approx 17 + \ln \left(\frac{b_{\mathrm{max}}}{10\,\mathrm{kpc}} \right) - \ln \left(\frac{M}{10^3\,M_\odot} \right) + 2\ln \left(\frac{|\vec{v}_M - \vec{v}_m|}{100\,\mathrm{km\,s}^{-1}} \right). \qquad (9.15)$$

However, unless the compact object is moving much more rapidly than the stellar velocity dispersion σ, it is safe to make the approximation that a typical encounter velocity will be $|\vec{v}_M - \vec{v}_m| \sim \sigma$. With this approximation, we can write

$$\ln \Lambda \approx 17 + \ln \left(\frac{R_{\mathrm{sys}}}{10\,\mathrm{kpc}} \right) - \ln \left(\frac{M}{10^3\,M_\odot} \right) + 2\ln \left(\frac{\sigma}{100\,\mathrm{km\,s}^{-1}} \right), \qquad (9.16)$$

with R_{sys} being the size of the stellar system through which the compact object is moving.[3]

When we assume a constant value of $\ln \Lambda$, we look back at Equation 9.14 and realize that the deceleration from dynamical friction follows an inverse square law. Admittedly, it's an inverse square law in *velocity* space rather than position space, but we can still use all the inverse-square-law tricks developed by Newton, including the shell theorems. If the velocity distribution $f_v(\vec{v}_m)$ is isotropic, and thus can be written as a distribution of speeds, $f(v_m)$, the frictional deceleration is

$$\frac{dv_M}{dt} = -16\pi^2 G^2 \rho_\star M \ln \Lambda \frac{1}{v_M^2} \int_0^{v_M} f(v_m) v_m^2 \, dv_m. \qquad (9.17)$$

Equation 9.17 is the Chandrasekhar dynamical friction formula, and applies to any isotropic distribution of stellar velocities. Newton's second shell theorem, in this application, implies that the compact object feels no net deceleration from stars with $v_m > v_M$. If the stars have a Maxwellian distribution of speeds (Equation 8.15), then the deceleration from dynamical friction can be written as

$$\frac{dv_M}{dt} = -\frac{4\pi G^2 \rho_\star M \ln \Lambda}{\sigma^2} G_{\mathrm{df}}(X). \qquad (9.18)$$

In Equation 9.18, the dependence on the compact object's speed v_M is given by the function

$$G_{\mathrm{df}}(X) = \frac{1}{2X^2} \left[\mathrm{erf}(X) - \frac{2}{\sqrt{\pi}} X \exp(-X^2) \right], \qquad (9.19)$$

where $\mathrm{erf}(X)$ is the error function and

$$X \equiv \frac{v_M}{\sqrt{2}\sigma}. \qquad (9.20)$$

---

[3] If we needed a factor $\exp(\Lambda)$, our rough approximations in computing Λ would be alarming. However, since we need $\ln \Lambda$, we are not alarmed.

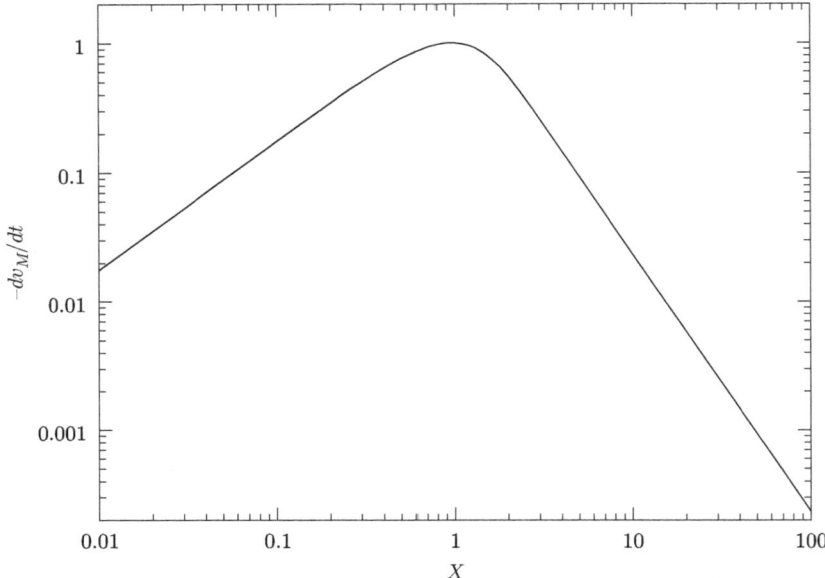

Figure 9.3 Dependence of the dynamical friction deceleration (from Equation 9.18) on the dimensionless speed $X \equiv v_M/(\sqrt{2}\sigma)$. Deceleration is normalized to its maximum value, which occurs at $X = 1$.

Figure 9.3 shows the magnitude of the deceleration, normalized to its maximum value at $X = 1$, where $G_{\mathrm{df}} \approx 0.214$.

If the compact object is very slow or very fast (compared to the stellar velocity dispersion σ), we may use the limiting cases

$$G_{\mathrm{df}}(X) \approx \begin{cases} 0.376\,X & [X \ll 1], \\ 0.500\,X^{-2} & [X \gg 1]. \end{cases} \tag{9.21}$$

Thus, for a slowly moving compact mass,

$$\frac{dv_M}{dt} \propto -v_M \qquad [v_M \ll \sigma]. \tag{9.22}$$

This represents a Stokes' law drag, just as you would see for a slowly moving object moving through a viscous fluid. By contrast, for a rapidly moving compact mass,

$$\frac{dv_M}{dt} \propto -\frac{1}{v_M^2} \qquad [v_M \gg \sigma]. \tag{9.23}$$

The "braking time,"

$$t_{\mathrm{brake}} \equiv v_M \left| \frac{dv_M}{dt} \right|^{-1}, \tag{9.24}$$

is thus independent of the compact object's speed in the slowly moving limit, but has the strong dependence $t_{brake} \propto v_M^3$ in the rapidly moving limit. At any speed, $t_{brake} \propto 1/M$; more massive compact objects are more effective at heating the stellar system through which they pass, and thus decelerate more rapidly.

Let's consider an example to see when dynamical friction is important in real stellar systems. Suppose that our compact mass M is a globular cluster orbiting within the dark halo of our galaxy. The mass distribution in the halo can be approximated as a singular isothermal sphere, with (Equation 6.97)

$$\rho(r) = \frac{v_{circ}^2}{4\pi G} \frac{1}{r^2}. \tag{9.25}$$

The isotropic velocity dispersion within the halo is related to the circular speed by the relation $\sigma^2 = v_{circ}^2/2$. If the globular cluster starts on a circular orbit, then initially $v_M = v_{circ}$ and $X \equiv v_M/(\sqrt{2}\sigma) = 1$. The dynamical friction force acting on the globular cluster is then, using Equation 9.18,

$$F_{df} = M\frac{dv_M}{dt} = -\frac{4\pi G^2 \rho M^2 \ln \Lambda}{\sigma^2} G_{df}(1), \tag{9.26}$$

which can be simplified, using Equation 9.25, into

$$F_{df} \approx -\frac{2GM^2 \ln \Lambda}{r^2} G_{df}(1) \approx -0.428 \ln \Lambda \frac{GM^2}{r^2}. \tag{9.27}$$

Since this force is always opposed to the globular cluster's direction of motion, the cluster will lose orbital angular momentum and spiral in toward the center of the halo.

We can compare the frictional force of Equation 9.27 to the gravitational force keeping the globular cluster on its orbit:

$$F_{grav} = -\frac{GMM_{halo}(r)}{r^2}, \tag{9.28}$$

where $M_{halo} = v_{circ}^2 r/G$ is the enclosed halo mass. Comparing Equations 9.27 and 9.28, we see that when $M_{halo} \gg M \ln \Lambda$, the frictional force will be much smaller than the gravitational force. Under these circumstances, the orbit of the cluster is *almost* circular, with specific angular momentum $j = v_{circ}r$.

The rate of loss of orbital angular momentum is given by

$$\frac{dj}{dt} = \frac{F_{df}}{M}r \approx -0.428 \ln \Lambda \frac{GM}{r}, \tag{9.29}$$

making use of Equation 9.27 for the dynamical friction force. Since $j = v_{circ}r$, with constant v_{circ} in an isothermal halo, this implies the radius of the cluster's orbit decreases at a rate

$$\frac{dr}{dt} = -0.428 \ln \Lambda \frac{GM}{v_{circ}} \frac{1}{r(t)}. \tag{9.30}$$

Adopting the initial condition $r = r_0$ at $t = 0$, and assuming a constant value for $\ln \Lambda$, this differential equation has the solution

$$r(t) = r_0 \left(1 - t/t_{df}\right)^{1/2}, \tag{9.31}$$

where

$$t_{df} = \frac{r_0^2 v_{circ}}{0.656 \, GM \ln \Lambda} \tag{9.32}$$

Scaled to properties of our galaxy's halo, the time required for a globular cluster to spiral inward is then

$$t_{df} \approx 6.2 \, \text{Gyr} \left(\frac{r_0}{1 \, \text{kpc}}\right)^2 \left(\frac{v_{circ}}{235 \, \text{km s}^{-1}}\right) \left(\frac{\ln \Lambda}{10}\right)^{-1} \left(\frac{M}{10^6 \, M_\odot}\right)^{-1}. \tag{9.33}$$

A massive globular cluster would have to start within roughly a kiloparsec of the galactic center in order to complete its death spiral during the $t_{sys} \sim 10 \, \text{Gyr}$ lifetime of our galaxy.

Equation 9.30 tells us that the speed with which the cluster drifts inward is $dr/dt \propto r(t)^{-1}$. Thus, during one orbital period, $\mathcal{P} = 2\pi r(t)/v_{circ}$, the cluster moves inward by a distance Δr that is independent of time. Using Equation 9.30, this distance is

$$\Delta r = \frac{dr}{dt} \frac{2\pi r}{v_{circ}} = -2.69 \ln \Lambda \frac{GM}{v_{circ}^2}. \tag{9.34}$$

Scaled to the properties of our galaxy, we find

$$\Delta r \approx 2.1 \, \text{pc} \left(\frac{\ln \Lambda}{10}\right) \left(\frac{M}{10^6 \, M_\odot}\right) \left(\frac{v_{circ}}{235 \, \text{km s}^{-1}}\right)^{-2}. \tag{9.35}$$

The trajectory of the globular cluster, as it spirals inward, is an Archimedean (or linear) spiral, with

$$\theta(r) = 2\pi \frac{r}{\Delta r} + \theta_0. \tag{9.36}$$

Figure 9.4 illustrates an Archimedean spiral. Unlike the logarithmic spiral of Figure 8.4, the Archimedean spiral traced out by the infalling cluster has a pitch angle that becomes larger as r decreases, with (Equation 8.58)

$$\tan \psi_p(r) = \frac{1}{r} \left| \frac{d\theta}{dr} \right|^{-1} = \frac{\Delta r}{2\pi r}. \tag{9.37}$$

Most globular clusters associated with our galaxy have orbits with low angular momentum, so their orbital evolution doesn't produce a perfect Archimedean spiral. However, the spiral approximation is useful for globular clusters such as NGC 6352, whose radial velocity and proper motion are consistent with a nearly circular orbit. The mean galactocentric distance $r \approx 3.6 \, \text{kpc}$ for NGC 6352 implies an orbital period $\mathcal{P} \approx 2\pi r/v_{circ} \approx 94 \, \text{Myr}$.

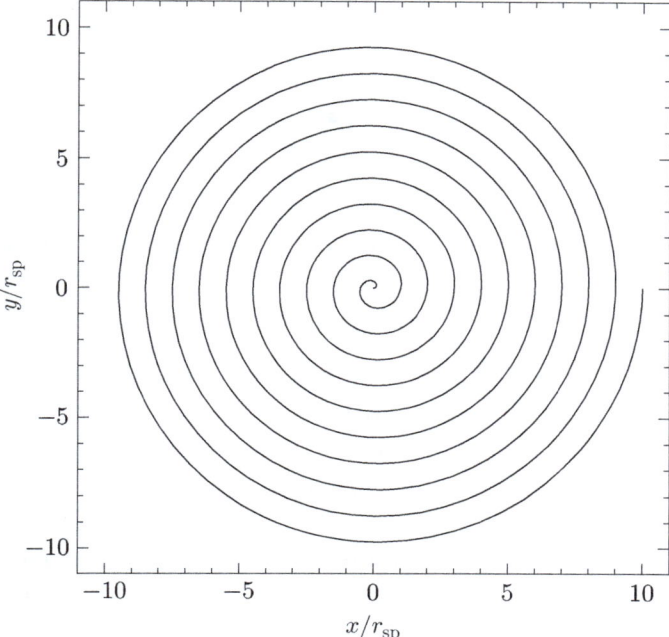

Figure 9.4 Archimedean "death spiral" of a compact object moving on a nearly circular orbit in a singular isothermal halo.

Since the cluster's age is $t_{\text{sys}} \approx 12.7\,\text{Gyr}$, it is a supercentenarian in Galactic years, with $t_{\text{sys}}/\mathcal{P} \approx 135$. The mass of NGC 6352 is $M \approx 6 \times 10^4\,\text{M}_\odot$, so during a single orbit this relatively lightweight globular cluster has $\Delta r \approx 0.13\,\text{pc}$, from Equation 9.35. During its lifetime, NGC 6352 has thus drifted inward by a distance of $\sim 135 \Delta r \sim 18\,\text{pc} \sim 0.005 r$.

9.2 High-Speed Encounters

Many galaxies are in bound clusters of galaxies. Using galaxies as luminous test masses, the one-dimensional velocity dispersion of a rich cluster is found to be $\sigma_{\text{clus}} \sim 1200\,\text{km s}^{-1}$. By comparison, inside a galaxy, using stars as luminous test masses, the one-dimensional velocity dispersion for an elliptical galaxy ranges from $\sigma_{\text{gal}} \sim 50\,\text{km s}^{-1}$ for a dwarf elliptical to $\sim 300\,\text{km s}^{-1}$ for a giant elliptical. Since $\sigma_{\text{gal}} < \sigma_{\text{clus}}$, it is possible for galaxies in a cluster to have a **high-speed encounter**, defined as an encounter for which the initial relative speed v is greater than the internal velocity dispersion of at least one of the two galaxies involved.

A useful approach to studying high-speed encounters involves using the **impulse approximation**. Suppose there exists a pair of spherical galaxies, as

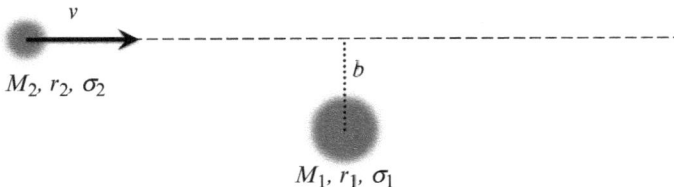

Figure 9.5 An encounter of two spherical stellar systems, with initial relative speed v and impact parameter b.

illustrated in Figure 9.5. Let's consider the effect that galaxy 2 (the "perturber") has on the internal structure of galaxy 1 (the "perturbee").[4] Galaxy 1 has a mass M_1, a characteristic radius r_1, and a velocity dispersion σ_1; the characteristic radius could be the truncation radius r_t, if the galaxy has fairly well-determined boundaries, or the half-mass radius r_h. Similarly, galaxy 2 has a mass M_2, a characteristic radius r_2, and a velocity dispersion σ_2. When the two galaxies are widely separated, their relative speed is v; the impact parameter of their encounter is b (Figure 9.5).

The gravitational force between the two galaxies is greatest when they are at their point of closest approach. We can define an effective encounter time t_{enc} as the length of time when the mutual gravitational force between the two galaxies is more than half its maximum value. For a high-speed encounter between two point masses,

$$t_{enc} \approx \frac{2b}{v}. \tag{9.38}$$

However, we must take into account the fact that galaxies are of finite size, and thus may overlap at the time of closest approach. For galaxies with radii r_1 and r_2, the effective encounter time is

$$t_{enc} \sim 2\frac{\max(b, r_1, r_2)}{v}. \tag{9.39}$$

In the impulse approximation, the force exerted on galaxy 1 by galaxy 2 (and vice versa) can be approximated as a single short, sharp shock at the time of closest approach, producing an instantaneous change in velocity for each star in galaxy 1, with no corresponding instantaneous change in position.

For the impulse approximation to hold true, the effective encounter time t_{enc} must be much shorter than the dynamical time of the perturbed galaxy. That is, we require

$$t_{enc} \ll \frac{r_1^3}{GM_1} \sim \frac{r_1}{\sigma_1}. \tag{9.40}$$

---

[4] Since gravity is a mutual force, we can find the effect on galaxy 2 from galaxy 1 simply by switching subscripts.

Thus, to use the impulse approximation safely, we must modify our definition of "high-speed encounter" to require

$$v \gg 2\sigma_1 \frac{\max(b, r_1, r_2)}{r_1}, \tag{9.41}$$

where the "1" subscript designates the galaxy being perturbed, and the "2" subscript designates the galaxy doing the perturbing.

Suppose that the perturbed galaxy consists of N stars, each with mass m_i, so that

$$M_1 = \sum_{i=1}^{N} m_i. \tag{9.42}$$

During the fast encounter, each star receives an impulsive change in velocity, $\Delta \vec{v}_i$. Since the perturbed galaxy is of finite size, different stars have different values of $\Delta \vec{v}_i$, with stars passing closer to the center of the perturber receiving larger "kicks." The barycenter of the perturbed galaxy receives a change of velocity

$$\Delta \vec{v}_{\text{bary}} = \sum_{i=1}^{N} m_i \Delta \vec{v}_i. \tag{9.43}$$

Relative to the galaxy's barycenter, each individual star has a change in velocity that is usually non-zero:

$$\Delta \vec{u}_i \equiv \Delta \vec{v}_i - \Delta \vec{v}_{\text{bary}}. \tag{9.44}$$

In general, computing $\Delta \vec{u}_i$ for each star in galaxy 1 requires knowing the gravitational potential of its perturber, galaxy 2. However, we can make some statements about the changes to the perturbed galaxy that are true regardless of the structure of the perturbing galaxy.

For instance, in the impulse approximation, the perturbed galaxy always has an *increase* in its internal energy E_1; the kinetic energy K_1 of stellar motions relative to the barycenter is increased, while the potential energy W_1 is unchanged in the impulse approximation. We can write

$$\Delta E_1 = \Delta K_1 = \frac{1}{2} \sum_{i=1}^{N} m_i |\Delta \vec{u}_i|^2 > 0. \tag{9.45}$$

The increase in the galaxy's internal energy comes from the kinetic energy of the relative motion of perturber and perturbee. In the impulse approximation for high-speed encounters, $M_1 v^2 / 2 \gg M_1 \sigma^2 / 2 \gg \Delta E_1$.

Prior to the encounter with galaxy 2, we expect that galaxy 1 was in virial equilibrium, with $K_1 = -E_1$ and $W_1 = -2K_1 = 2E_1$. However, immediately after the impulsive encounter, galaxy 1 is thrown out of virial equilibrium, with

$$K_1' = -E_1 + \Delta E_1, \tag{9.46}$$

$$W_1' = 2E_1, \tag{9.47}$$

$$E_1' = E_1 - \Delta E_1. \tag{9.48}$$

After approximately one dynamical time ($t_{dyn} \sim R_1/\sigma_1$), galaxy 1 will return to virial equilibrium, with new equilibrium energies

$$K_1'' = -E_1 - \Delta E_1, \tag{9.49}$$

$$W_1'' = 2E_1 + 2\Delta E_1, \tag{9.50}$$

$$E_1'' = E_1 + \Delta E_1. \tag{9.51}$$

At first, the high-speed encounter increases the kinetic energy by an amount $\Delta E_1 > 0$; however, it ends by decreasing the kinetic energy by an amount $-\Delta E_1$ from its initial value.

In general, computing the value of ΔE_1 for the perturbed galaxy requires detailed knowledge of the potential of the perturbing galaxy. However, there are some limiting cases where we can make simplifying assumptions. As an example, consider the case of **head-on** high-speed encounters, for which $b = 0$. If the perturbing object is a compact mass with $r_2 \ll r_1$, then the duration of the encounter will be $t_{enc} \sim r_1/v$. The typical acceleration felt by a star in the large perturbed galaxy during the encounter will be $g \sim GM_2/r_1^2$. Thus, stars in the perturbed galaxy will receive velocity kicks of order

$$\Delta u_1 \sim g t_{enc} \sim \frac{GM_2}{r_1 v} \qquad [r_1 \gg r_2]. \tag{9.52}$$

By symmetry, a head-on encounter with a dense "bullet" will not change the speed of the perturbed galaxy's barycenter. The change in energy of the perturbed galaxy, in our order-of-magnitude estimate, will be

$$\Delta E_1 \sim M_1 (\Delta u_1)^2 \sim \left(\frac{GM_2}{r_1 v} \right)^2 M_1 \qquad [r_1 \gg r_2]. \tag{9.53}$$

Since the internal energy of the perturbed galaxy is $E_1 \sim -GM_1^2/r_1$, the fractional change in energy from a high-speed head-on encounter is

$$\frac{\Delta E_1}{|E_1|} \sim \frac{GM_1}{r_1} \frac{1}{v^2} \frac{M_2^2}{M_1^2} \sim \left(\frac{\sigma_1}{v} \right)^2 \left(\frac{M_2}{M_1} \right)^2 \qquad [r_1 \gg r_2]. \tag{9.54}$$

Since $\sigma_1 < v$ in our high-speed approximation, we conclude that a high-speed "bullet" can disrupt a galaxy only if it has a mass greater than that of the perturbed galaxy, with $M_2 > (v/\sigma_1)M_1$. A lower-mass bullet will not be able

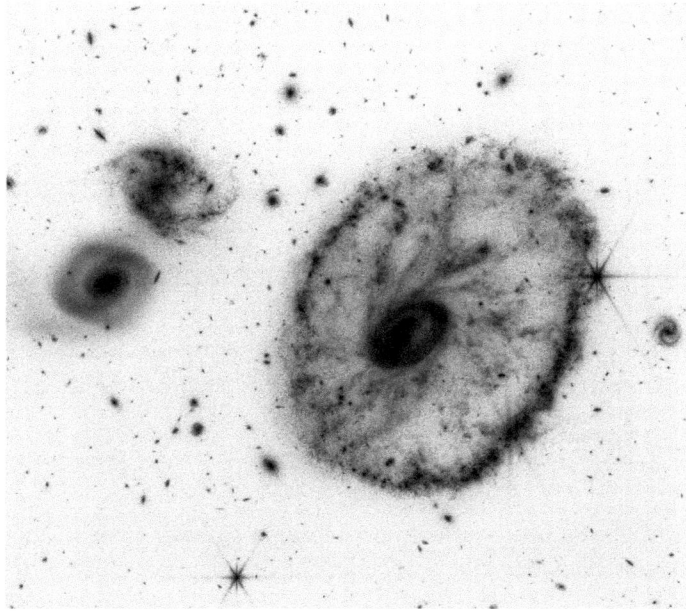

Figure 9.6 Near-infrared image of the Cartwheel Galaxy, at a distance $d \sim 130\,\mathrm{Mpc}$. The image size is $2.34 \times 2.15\,\mathrm{arcmin}$, corresponding to $\sim 88 \times 81\,\mathrm{kpc}$ at the distance of the Cartwheel Galaxy. [NASA, ESA, CSA, STScI, Webb ERO Production Team]

to completely disrupt the perturbed galaxy. However, it will be able to produce axisymmetric density waves, like those seen in the Cartwheel Galaxy (Figure 9.6). The perturbing "bullet" is probably not either of the two small companion galaxies seen in Figure 9.6; the most likely suspect is already at a projected separation $r \sim 120\,\mathrm{kpc}$ from the center of the Cartwheel Galaxy, and is thus outside the field of view of Figure 9.6. The nearly head-on encounter occurred $\sim 300\,\mathrm{Myr}$ ago, indicating that the relative speed was at least $\sim 0.4\,\mathrm{kpc}\,\mathrm{Myr}^{-1} \sim 400\,\mathrm{km}\,\mathrm{s}^{-1}$.

So far, in our discussion of head-on encounters, we have assumed that the perturber (galaxy 2) is the smaller galaxy, while the perturbee (galaxy 1) is the larger galaxy. We can, however, switch the labels so that the larger galaxy is the perturber (galaxy 2) and the smaller compact galaxy is the perturbee (galaxy 1). In the limit $r_2 \gg r_1$, the duration of the encounter will be $t_{\mathrm{enc}} \sim r_2/v$. The acceleration of each star in the compact galaxy will be nearly the same; however, there will be a differential tidal force across the compact galaxy, giving a typical acceleration

$$g \sim \frac{GM_2 r_1}{r_2^3} \tag{9.55}$$

to stars in the compact galaxy. The typical velocity kick given to each star in the compact galaxy will be

$$\Delta u_1 \sim g t_{\mathrm{enc}} \sim \frac{G M_2 r_1}{v r_2^2} \qquad [r_1 \ll r_2]. \qquad (9.56)$$

The change of internal energy for the perturbed compact galaxy will be

$$\Delta E_1 \sim M_1 (\Delta u_1)^2 \sim M_1 \left(\frac{G M_2 r_1}{v r_2^2} \right)^2 \qquad [r_1 \ll r_2]. \qquad (9.57)$$

Since the internal energy of the perturbed compact galaxy is $E_1 \sim -G M_1^2 / r_1$, the fractional change in its energy from a high-speed head-on encounter is

$$\frac{\Delta E_1}{|E_1|} \sim \frac{G M_1}{r_1} \frac{1}{v^2} \frac{M_2^2}{M_1^2} \frac{r_1^4}{r_2^4} \sim \left(\frac{\sigma_1}{v} \right)^2 \left(\frac{M_2}{M_1} \right)^2 \left(\frac{r_1}{r_2} \right)^4 \qquad [r_1 \ll r_2]. \qquad (9.58)$$

As an example of a large object perturbing a smaller system during a high-speed encounter, consider the perturbation of an open cluster of stars by a giant molecular cloud within our galaxy. (A giant molecular cloud, of course, is not a stellar system. However, since we are taking the open cluster to be the perturbed system, it doesn't matter what the perturber is made of, as long as it has a non-zero mass M_2.) Most open clusters are quite young compared to the age of our galaxy; the median age of open clusters is $t_{\mathrm{sys}} \sim 100 \, \mathrm{Myr}$, with only $\sim 10\%$ of open clusters being more than 1 Gyr old. Most open clusters are also close to the midplane of our galaxy's disk, with a scale height $h_z \sim 100 \, \mathrm{pc}$ for the distribution of open clusters. Since giant molecular clouds are also close to the midplane (with $h_z \sim 75 \, \mathrm{pc}$ for molecular clouds), this makes an encounter between an open cluster and a giant molecular cloud more likely than if they were more spread out in space. Since open clusters and molecular clouds are on relatively orderly, nearly circular orbits about the galactic center, the relative velocity of their encounters will be small: let's take $v = 10 \, \mathrm{km \, s^{-1}}$ as a typical encounter speed.

As our example of an open cluster, we take the nearby Hyades. At a distance of only $d \approx 47 \, \mathrm{pc}$, the Hyades has a particularly well-observed structure. Its mass is $M_1 \approx 400 \, \mathrm{M_\odot}$ and its half-mass radius is $r_1 \approx 5.7 \, \mathrm{pc}$; the velocity dispersion in its central regions is $\sigma_1 \approx 0.3 \, \mathrm{km \, s^{-1}}$. Molecular clouds with masses in the range $10^4 \, \mathrm{M_\odot} < M_2 < 10^7 \, \mathrm{M_\odot}$ are usually granted the title of "giant" molecular clouds. The observed mass–radius relation for giant molecular clouds is

$$r_2 \sim 30 \, \mathrm{pc} \left(\frac{M_2}{10^6 \, \mathrm{M_\odot}} \right)^{1/2}. \qquad (9.59)$$

In an encounter with a high-mass cloud, the open cluster will then be the smaller of the two objects. A nearly head-on encounter ($b < r_2 \sim 30 \, \mathrm{pc}$)

between an open cluster and a giant molecular cloud will increase the internal energy of the open cluster by a fractional amount (Equation 9.58)

$$\frac{\Delta E_1}{|E_1|} \sim \left(\frac{\sigma_1}{v}\right)^2 \left(\frac{M_2}{M_1}\right)^2 \left(\frac{r_1}{r_2}\right)^4 \sim 7 \left(\frac{v}{10\,\mathrm{km\,s^{-1}}}\right)^{-2}, \qquad (9.60)$$

assuming the open cluster has the same properties as the Hyades. (Notice that since $M_2 \propto r_2^2$ for the molecular cloud, the mass of the molecular cloud doesn't really matter in Equation 9.60.) Although this is a rough order-of-magnitude calculation, it does indicate that a low-speed encounter with a giant molecular cloud is capable of disrupting an open cluster. The small number of open clusters observed to be older than ~3 Gyr tend to be on orbits that keep them away from the cluster-busting molecular clouds in the midplane of our galaxy. The oldest known open cluster, NGC 6791, has an age $t_{\mathrm{sys}} \sim 8$ Gyr, but is currently at a distance $z \sim 800$ pc from the midplane of our galaxy.

9.3 Distant Encounters

Since stellar systems have a finite size, they can be significantly affected by the differential tidal force exerted by neighboring massive objects. To begin with a simple case, consider a high-speed encounter, like that illustrated in Figure 9.5, but with an impact parameter restricted to be $b \gg r_1 + r_2$. That is, even at closest approach, the two galaxies are far from overlapping. Suppose that the perturber can be approximated as a compact object with mass M_2. As it zips past, it gives each star in the perturbed galaxy a different velocity kick $\Delta \vec{u}_i$ relative to the barycenter of the perturbed galaxy. For convenience, let's choose a coordinate system in which the barycenter of the perturbed galaxy is at the origin, the z axis lies in the direction of the initial mutual velocity $\vec{v}$, and the y axis points from the perturbed galaxy to the perturber at the time of closest approach. By symmetry about the time of closest approach, we don't expect a net kick $\Delta v_{z,i}$ in the direction of mutual motion. In the other two directions, each star receives a velocity change

$$\Delta \vec{u}_i = \frac{2GM_2}{b^2 v}(-x_i \hat{x} + y_i \hat{y}) \qquad (9.61)$$

relative to the barycenter of the perturbed galaxy. Thus, the velocity kick stretches the perturbed galaxy along the y axis and squeezes it along the x axis (perpendicular to the two-dimensional diagram of Figure 9.5). The perturbed galaxy can be thought of as receiving a single tidal flex when the perturber is at closest approach; this is in contrast to the moon Io, which is subjected to continuous tidal flexing on its orbit around Jupiter, and thus is continuously heated.

As a result of a single distant flyby by a perturber, the perturbed galaxy's internal energy is slightly increased. We can use Equation 9.45 to compute the change in energy:

$$\Delta E_1 = \sum_i m_i |\Delta \vec{u}_i|^2 = \frac{2G^2 M_2^2}{b^4 v^2} \sum_i m_i (x_i^2 + y_i^2). \qquad (9.62)$$

For an initially spherical galaxy, with $\langle x^2 \rangle = \langle y^2 \rangle = \langle z^2 \rangle = \langle r^2 \rangle / 3$, the energy increase can be written as

$$\Delta E_1 = \frac{4G^2 M_2^2 M_1}{3 b^4 v^2} \langle r_1^2 \rangle, \qquad (9.63)$$

where $\langle r_1^2 \rangle$ is the mean square radius of the perturbed galaxy, weighted by stellar mass. Jettisoning all factors of order unity, the fractional change in a galaxy's internal energy from a single distant flyby is

$$\frac{\Delta E_1}{E_1} \sim \left(\frac{\sigma_1}{v} \right)^2 \left(\frac{M_2}{M_1} \right)^2 \left(\frac{r_1}{b} \right)^4. \qquad (9.64)$$

This is smaller by a factor $(r_1/b)^4$ than it would be if the compact perturber flew straight through the center of the perturbed galaxy (compare to Equation 9.54).

9.4 Tidal Stripping

Two stellar systems that have a single high-speed encounter will feel a single tidal flex. However, there are situations where the tidal distortion of a stellar system is continuous, and has a significant effect on the system's overall structure. Consider, for instance, a star cluster or dwarf galaxy orbiting in the extended halo of a much larger galaxy. The smaller stellar structure will then feel the effects of **tidal stripping**. The concept of tidal stripping is related to the Hill radius, as discussed in Section 2.2. If two point masses, with $m_1 > m_2$, orbit their barycenter on a circular orbit with radius a, then the secondary has a Hill radius given by the relation (Equation 2.34)

$$r_H \approx r_{pe} \left(\frac{m_2}{3 m_1} \right)^{1/3}, \qquad (9.65)$$

where r_{pe} is the periapsis distance for the two-body system. Test masses that stray outside the Hill radius cannot remain bound to the secondary. If we assume the secondary is a star cluster or dwarf galaxy of finite size, then we expect its radius r_2 must satisfy the relation $r_2 < r_H$; any stars beyond the Hill radius will be tidally stripped. As a rough first estimate, we can make the "point halo" approximation, and state that the Hill radius of a globular cluster orbiting in a galaxy's halo is given by Equation 9.65; we simply replace m_1 for the point mass primary with the enclosed mass of the halo (Equation 6.98),

$$M(r) = \frac{v_{\text{circ}}^2}{G} \approx 10^{11}\,\text{M}_\odot \left(\frac{v_{\text{circ}}}{235\,\text{km s}^{-1}}\right)^2 \left(\frac{r_{\text{pe}}}{8.2\,\text{kpc}}\right). \tag{9.66}$$

Scaled to a medium-sized globular cluster ($m_2 = M_{\text{gc}} \approx 10^5\,\text{M}_\odot$) in our galaxy's halo, we find a Hill radius (also called a "tidal radius" in this context)

$$r_{\text{H}} = \left(\frac{GM_{\text{gc}}r_{\text{pe}}^2}{3v_{\text{circ}}^2}\right)^{1/3} \tag{9.67}$$

$$\approx 56\,\text{pc} \left(\frac{M_{\text{gc}}}{10^5\,\text{M}_\odot}\right)^{1/3} \left(\frac{v_{\text{circ}}}{235\,\text{km s}^{-1}}\right)^{-2/3} \left(\frac{r_{\text{pe}}}{8.2\,\text{kpc}}\right)^{2/3}.$$

Note that if a globular cluster fully occupies its Hill radius, its mean density is independent of the cluster mass, with

$$\overline{\rho}_{\text{gc}} = \frac{3M_{\text{gc}}}{4\pi r_{\text{H}}^3} = \frac{9v_{\text{circ}}^2}{4\pi Gr_{\text{pe}}^2}. \tag{9.68}$$

This yields a dynamical time for the cluster of

$$t_{\text{dyn,gc}} \sim (G\overline{\rho}_{\text{gc}})^{-1/2} \sim 5\,\text{Myr} \left(\frac{v_{\text{circ}}}{235\,\text{km s}^{-1}}\right)^{-1} \left(\frac{r_{\text{pe}}}{1\,\text{kpc}}\right). \tag{9.69}$$

For the globular clusters in our galaxy's halo, this dynamical time is much shorter than the typical age $t_{\text{sys}} \sim 10\,\text{Gyr}$ of a globular cluster.

Although the "point halo" approximation that went into Equation 9.67 is admittedly crude, it is found observationally that globular clusters do have a fairly abrupt truncation radius, at which the surface brightness of the cluster drops rapidly. The King models discussed in Section 8.1.1 were partially inspired by the observed truncation radii of globular clusters in our galaxy. From the observed proper motions and radial velocities of globular clusters, we can get a fairly good idea of their perigalactic distance r_{pe}; that is, their distance of closest approach to the galactic center. From the internal motions of stars within a globular cluster, we can make an estimate of the cluster mass M_{gc}; combined with the value for r_{pe}, this permits us to compute r_{H}.

Figure 9.7 compares the fitted truncation radius r_{t} to the computed Hill radius r_{H} for a sample of 80 globular clusters orbiting in our galaxy's halo. On average, the truncation radius equals the Hill radius for globular clusters. However, there is considerable scatter. Some of the scatter is due to uncertainties in the values found for r_t, r_{pe}, and M_{gc}. However, some of the scatter points to interesting physics. Consider, for instance, the conspicuous outlier in Figure 9.7, labeled "Pal 14." This is the globular cluster Palomar 14, which has a fitted tidal radius $r_{\text{t}} \approx 5r_{\text{H}}$. The spread of Pal 14 beyond its computed

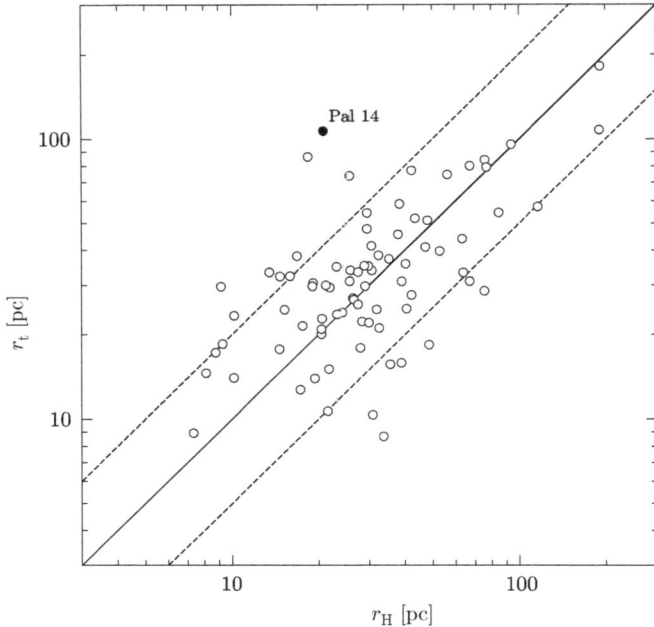

Figure 9.7 Fitted truncation radius r_t of a King model versus computed Hill radius r_H for a sample of 80 globular clusters. The diagonal solid line is $r_t = r_H$, while the dashed lines are $r_t = r_H/2$ and $r_t = 2r_H$. [Data from Harris 1996, Baumgardt and Hilker 2018, Baumgardt *et al.* 2019]

Hill radius is related to the fact that it is on a large, nearly radial orbit. It is currently at a distance $r \approx 68\,\mathrm{kpc}$ from the galactic center; it passed through perigalacticon, at $r_{pe} \approx 4\,\mathrm{kpc}$, about 230 Myr ago, and is now en route to apogalacticon at $r_{ap} \approx 95\,\mathrm{kpc}$. During its most recent perigalactic passage, Pal 14 was tidally stripped during a relatively brief encounter that threw it out of virial equilibrium. "Relatively brief," in this case, means that most of the stripping took place on a timescale shorter than the dynamical time of the cluster, $t_{\mathrm{dyn,gc}} \sim 20\,\mathrm{Myr}$. However, the time that has elapsed since perigalacticon, $t \sim 230\,\mathrm{Myr}$, is longer than the cluster's dynamical time. This means that Pal 14 has had time to restore virial equilibrium, a process that involves an expansion in its radius. In addition, the stars that are stripped from the globular cluster do not disappear; instead, they follow galactocentric orbits that keep them, at least initially, in close proximity to the cluster. With time, the distance between the stripped stars and the cluster center grows, creating long **tidal tails**. Detailed analysis of Pal 14, for instance, reveals tidal tails that are ~1° long, equivalent to ~1200 pc at the distance of Pal 14.

   On close examination, most globular clusters and dwarf galaxies orbiting in our galaxy's halo display tidal tails. The globular cluster Palomar 5 has

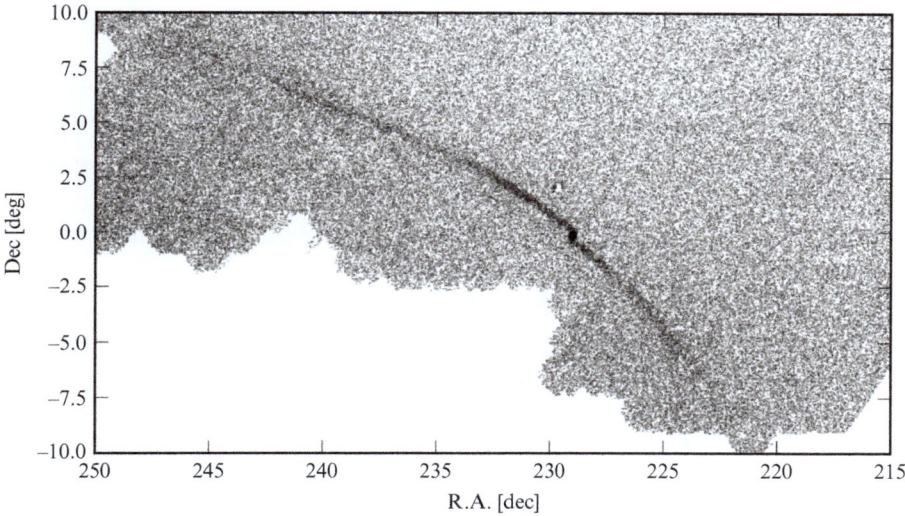

Figure 9.8 Tidal tails of the globular cluster Palomar 5, at a distance $d \approx 23\,\mathrm{kpc}$. [Bonaca *et al.* 2020]

particularly prominent and massive tidal tails, shown in Figure 9.8. The leading tail (to the lower right of Figure 9.8) can be traced over ∼10° in angle; the trailing tail can be traced over ∼15°. At the distance of Pal 5, this means that the tails form a coherent structure over a length ∼10 000 pc. The total mass of the tails is comparable to the mass $M_{\mathrm{gc}} \approx 1.4 \times 10^4\,M_\odot$ remaining in the globular cluster itself. However, the cluster's loss is astronomy's gain: the properties of tidal tails can be used as a probe of the distribution of dark matter in the halo. The overall shape of the tails gives information about the galactic potential, while smaller-scale kinks and gaps can be related to the presence of substructure within the halo.

Exercises

9.1 The Sun is surrounded by the **Oort cloud**, which can be roughly approximated as a spherical distribution of comets with number density

$$n(r) = n_{\mathrm{i}} \left(\frac{r}{r_{\mathrm{i}}} \right)^{-3} \tag{9.70}$$

between an inner radius $r_{\mathrm{i}} = 3000\,\mathrm{au}$ and an outer radius $r_{\mathrm{o}} = 10^5\,\mathrm{au}$.

(a) If there are 10^{12} comets in the Oort cloud, what is the value of n_{i} at its inner edge?

(b) If the mean cometary mass is equal to that of Comet Halley, $m = 2.2 \times 10^{17}\,\mathrm{g}$, what is the total mass M_{oort} of the Oort cloud? Give your answer in units of the Sun's mass.

(c) What is the gravitational potential energy W_{oort} of the Oort cloud?

9.2 About 1.3 Myr from now, the star Gliese 710 will make a relatively close passage by the Sun. The impact parameter for the encounter between the Sun and Gliese 710 is $b_\star = 10\,700$ au, with an initial relative speed $v_\star = 15\,\text{km s}^{-1}$. The mass of Gliese 710 is $M_G = 0.57\,M_\odot$.

(a) What is the initial kinetic energy K of the unbound binary system consisting of the Sun and Gliese 710?

(b) As Gliese 710 moves through the Oort cloud, what will be its critical impact parameter b_{cge} for close gravitational encounters with a comet? How many comets in the Oort cloud will undergo a close gravitational encounter with Gliese 710? [Use the Oort cloud properties from the previous problem.]

(c) Assuming a Maxwellian distribution of comet speeds, compute the rate dK/dt at which Gliese 710 will lose kinetic energy through dynamical friction as it passes through the Oort cloud. [*Hint:* Will star–comet encounters be at high speeds ($X \gg 1$) or low speeds ($X \ll 1$)?]

(d) Approximately how much energy ΔE will be transfered to the Oort cloud from Gliese 710 by dynamical friction? Compare this both to the initial kinetic energy of Gliese 710 and to the gravitational potential energy of the Oort cloud (computed in the previous problem).

10

Relaxation and Collisional Systems

I'm trying to relax, but it's hard.

Ryuichi Sakamoto (1952–2023)
New York Times interview, 2017 Apr 22

In the previous few chapters, we have been dealing mainly with **collision-less** stellar systems, where the behavior of individual stars can be predicted using the assumption that they are moving through a smooth density distribution $\rho(\vec{r}, t)$ that varies only gradually with time. However, as we found in the introduction to Chapter 6, occasionally a star undergoes a close gravitational encounter with another star. These close encounters cause **relaxation** of the stellar system. The word "relaxation" is derived from the Latin *laxus*, meaning loose, or slack. Physicists first used the word "relaxation" to mean the return to equilibrium of a system that has been subjected to a strain; eventually the term broadened its meaning to any process by which a system goes from a high-energy, low-entropy state to a lower-energy, higher-entropy state.

10.1 Relaxation

The type of relaxation that results from close gravitational encounters between stars is called **two-body relaxation**. To find the characteristic time for two-body relaxation, start by choosing an individual star within a stellar system. Initially, the velocity of the star is known to have a value $\vec{v}$. At some later time t, let $\vec{v}_0(t)$ be the predicted velocity of the star in the absence of close gravitational encounters, while $\vec{v}_1(t)$ is its true velocity, with gravitational encounters taken into account. The stellar system is **relaxed** when

$$\frac{|\vec{v}_0(t) - \vec{v}_1(t)|}{|\vec{v}_0(t)|} \approx 1. \tag{10.1}$$

That is, the system is relaxed when the approximation of a perfectly smooth density ρ starts to give a wretchedly bad approximation of a typical star's trajectory. Since a single close gravitational encounter is enough to completely change a star's velocity, we can approximate the **relaxation time** t_{rel} as being the mean time between close gravitational encounters. As given in Equation 6.13,

$$t_{rel} \approx t_{cge} \approx \frac{v^3}{4\pi n_\star (2Gm_\star)^2},$$
(10.2)

where v is a typical relative speed for stars in the system, $n_\star$ is the number density of stars, and $m_\star$ is the average mass of the two interacting stars.

Consider a spherical stellar system. Since the density $n_\star$ and velocity distribution $f(\vec{v})$ will vary with radius, the relaxation time will vary as well; there can be stellar systems that are relaxed in their dense central regions but not in their lower-density outskirts. Suppose that a fraction F of the system's stars are within a distance r_F of the center. (With this definition, choosing $F = 0.5$ gives you the half-mass radius r_h of the system.) The mean density of stars inside r_F is then

$$\bar{n}(r_F) = \frac{3}{4\pi} \frac{FN}{r_F^3},$$
(10.3)

where N is the total number of stars in the system. If the system has an isotropic velocity dispersion, then a typical relative velocity v will be the three-dimensional dispersion $\sqrt{3}\sigma$. If the system is in virial equilibrium, we expect the dispersion in the vicinity of r_F to be

$$3\sigma(r_F)^2 \sim \frac{Gm_\star(FN)}{r_F}.$$
(10.4)

Making the substitutions $n_\star = \bar{n}(r_F)$ and $v = \sqrt{3}\sigma(r_F)$ in Equation 10.2, we find a two-body relaxation time

$$t_{rel,F} \sim \frac{1}{12}\left(FN\frac{r_F^3}{Gm_\star}\right)^{1/2},$$
(10.5)

where the numerical factor in front should not be taken too seriously in this order-of-magnitude calculation. It is interesting to compare this relaxation time to the dynamical time at r_F, which we define as

$$t_{dyn,F} \equiv [Gm_\star\bar{n}(r_F)]^{-1/2} \approx 2.05\left(\frac{1}{FN}\frac{r_F^3}{Gm_\star}\right)^{1/2} \approx 1.18\frac{r_F}{\sigma(r_F)}.$$
(10.6)

Using Equations 10.5 and 10.6, we can write the two-body relaxation time at radius r_F in terms of the dynamical time at that radius:

$$t_{rel,F} \sim 0.04FNt_{dyn,F}.$$
(10.7)

A more careful calculation of the relaxation time must take into account the cumulative effect of distant encounters, in addition to the occasional close gravitational encounter. (Being nibbled away by minnows is a possible alternative to being gulped down by a shark.) The careful calculation, including all encounters, yields

$$t_{\text{rel},F} \approx 0.12 \frac{FN}{\ln(\lambda N)} t_{\text{dyn},F}. \tag{10.8}$$

The factor of $\ln(\lambda N)$ is a lightly disguised version of our old friend, the $\ln \Lambda$ term in the dynamical friction formula of Equation 9.13. For realistic stellar systems, adopting $\lambda \approx 0.2$ gives a good fit to the relaxation time. (Note that the dynamical time $t_{\text{dyn},F}$ depends only on the mass interior to r_F; however, two-body relaxation occurs via encounters with individual stars at all radii. Thus, the factor λ doesn't vary strongly with F.)

As an example of a stellar system where two-body relaxation is important, consider the Hyades. As mentioned in Section 9.2, this open cluster has a half-mass radius $r_{\text{h}} \approx 5.7\,\text{pc}$ and a velocity dispersion $\sigma \approx 0.3\,\text{km}\,\text{s}^{-1}$. The dynamical time at the half-mass radius is thus $t_{\text{dyn,h}} \sim r_{\text{h}}/\sigma \sim 20\,\text{Myr}$. There are $N \sim 1000$ stars in the Hyades, yielding a relaxation time $t_{\text{rel,h}} \sim 10 t_{\text{dyn,h}} \sim 200\,\text{Myr}$ at the half-mass radius. Since the age of the Hyades is $t_{\text{sys}} \approx 625\,\text{Myr}$, it is an example of a relaxed system, with $t_{\text{rel,h}} < t_{\text{sys}}$. The majority of the stars in the Hyades have "forgotten" what their original orbits were.

Although globular clusters have half-mass radii comparable to those of open clusters, they usually contain a much larger number N of stars. This means that globular clusters tend to have shorter dynamical times but longer relaxation times than open clusters. Consider, as an example, the nearby globular cluster NGC 6397, at a distance $d = 2.5\,\text{kpc}$. The half-mass radius of this cluster is $r_{\text{h}} \approx 4.0\,\text{pc}$ and its central velocity dispersion is $\sigma \approx 5.2\,\text{km}\,\text{s}^{-1}$. This gives a half-mass dynamical time of $t_{\text{dyn,h}} \sim 0.8\,\text{Myr}$, much shorter than that of the low-density Hyades. However, NGC 6397 contains $\sim 2 \times 10^5$ stars; this yields a half-mass relaxation time $t_{\text{rel,h}} \sim 1100 t_{\text{dyn,h}} \sim 900\,\text{Myr}$, significantly longer than that of the Hyades. Since NGC 6397 is an old stellar system, with $t_{\text{sys}} \approx 12\,\text{Gyr}$, it is nevertheless old enough for its central regions to be relaxed.

The process of two-body relaxation has a number of observable side effects on stellar systems. In general, close gravitational encounters increase the entropy of a stellar system. The random velocity kicks given to stars during gravitational encounters drive the velocity dispersion σ_{ij} toward isotropy, and the velocity distribution $f(v)$ toward a Maxwellian distribution (Equation 8.15). In addition, two-body relaxation drives the physical shape of a stellar system toward something spherical and smooth, with the destruction of substructure. If a stellar system contains a dense core and a tenuous outer envelope, entropy can be increased by making the core more compact and the outer envelope more extended.

Two-body relaxation also results in **equipartition** of kinetic energy. In the frame of reference of the stellar system's barycenter, a snapshot at a given time t will reveal that individual stars will have different kinetic energies. However, a close gravitational encounter between two stars of different kinetic energy will (on average) transfer kinetic energy from the higher-energy star to the lower-energy star. Thus, in a system that has undergone two-body relaxation, stars with different masses will have the same mean kinetic energy. For example, if we compare sun-like stars with $m_1 = 1 \, M_\odot$ to red dwarf stars with $m_2 = 0.1 \, M_\odot$, in a relaxed cluster we will find

$$\frac{1}{2} m_1 \langle v_1^2 \rangle = \frac{1}{2} m_2 \langle v_2^2 \rangle \tag{10.9}$$

and thus

$$\frac{\langle v_1^2 \rangle}{\langle v_2^2 \rangle} = \frac{m_2}{m_1}. \tag{10.10}$$

If the high-mass and low-mass stars in a cluster initially had the same spatial distribution and velocity distribution, then the transfer of kinetic energy would cause the high-mass stars to lose energy and sink toward the center of the cluster. This process is known as **mass segregation**.

The effects of mass segregation can be observed in globular clusters. The luminosity density $\Psi_\star(r)$ of a cluster can be determined by deprojecting its surface brightness; similarly, the number density of stars $n_\star(r)$ can be found by deprojecting number counts of stars. Since the luminosity of a main sequence star is strongly dependent on mass, the visible sign of mass segregation would be a mean stellar luminosity $\Psi_\star/n_\star$ that rises toward the center. In the cluster NGC 6397, which we have used as our example of a relaxed globular cluster, such a radial variation in mean stellar luminosity is detected. Within NGC 6397, the number density $n_\star(r)$ is well fitted by a King model with a well-resolved core radius $r_K = 42 \, \text{arcsec}$ (equal to $0.50 \, \text{pc}$ at the distance of NGC 6397). However, the luminosity density $\Psi_\star$ continues to increase as $r \to 0$, all the way to the limit of resolution. The resulting radial dependence of the mean luminosity per star is shown in Figure 10.1(a), expressed as the apparent magnitude in the V band. The difference of ~ 4 magnitudes between the stars in the innermost bin and stars in the outer regions translates to a factor of ~ 50 in luminosity. At the cluster's distance of $d = 2.5 \, \text{kpc}$, the mean V-band luminosity drops from $\overline{L}_V \approx 4 \, L_{V,\odot}$ near the center to $\overline{L}_V \approx 0.08 \, L_{V,\odot}$ in the cluster's outskirts. Given the great age of the cluster ($t_{\text{sys}} \approx 12 \, \text{Gyr}$), stars with $L_V \approx 4 \, L_{V,\odot}$ are not main sequence stars, but rather are stars with mass $m_\star \approx 0.8 \, M_\odot$ that are starting to swell into red giants. However, stars with $\overline{L}_V \approx 0.08 \, L_{V,\odot}$ are thrifty lower-mass stars, with $m_\star \approx 0.56 \, M_\odot$, that are

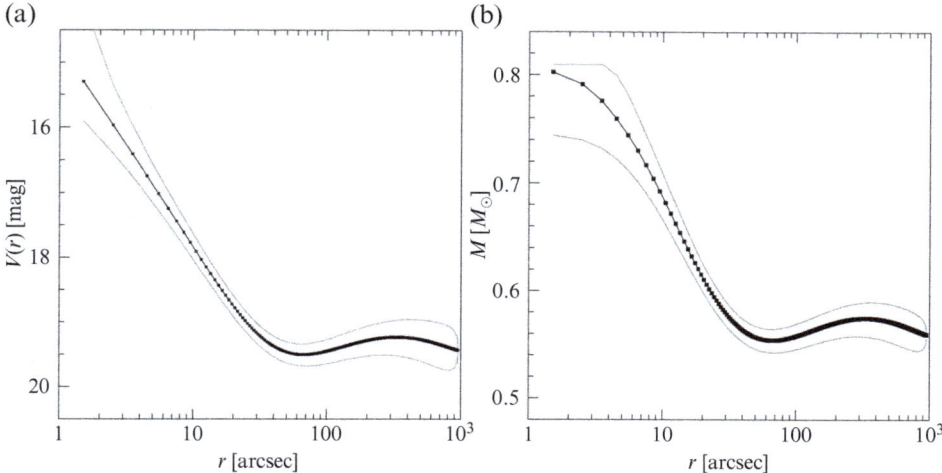

Figure 10.1 (a) Mean stellar luminosity (expressed as V-band apparent magnitude) as a function of radius in the globular cluster NGC 6397. (b) Stellar mass corresponding to the mean luminosity, assuming a stellar population with $t = 12\,\text{Gyr}$ and $[\text{Fe/H}] = -1.8$. [Martinazzi *et al.* 2014]

still fusing hydrogen into helium on the main sequence. Figure 10.1(b) provides a translation between the mean luminosity $\overline{L}_V$, shown in Figure 10.1(a), and the corresponding stellar mass.

Two-body relaxation also results in **evaporation** of stars. Gravitational encounters with other stars will cause a star to wander about in phase space; as a result, the star may eventually find that its speed is greater than the escape speed at its location. The escape speed from a location $\vec{r}$ is given by the relation (see Equation 6.80):

$$v_{\text{esc}}(\vec{r})^2 = -2\Phi(\vec{r}). \qquad (10.11)$$

The mass-weighted mean square escape speed for the entire stellar system is then

$$\langle v_{\text{esc}}^2 \rangle = -\frac{2}{M} \int \rho(\vec{r})\Phi(\vec{r})\, d^3r, \qquad (10.12)$$

where M is the total mass of the stellar system. However, Equation 6.58 reminds us that the gravitational potential energy of the system is

$$W = \frac{1}{2} \int \rho(\vec{r})\Phi(\vec{r})\, d^3r. \qquad (10.13)$$

Thus, the mean square escape speed can be simply related to the gravitational potential energy:

$$\langle v_{\text{esc}}^2 \rangle = -\frac{4W}{M}. \qquad (10.14)$$

If the stellar system is in virial equilibrium, we can use the relation $W = -2K$ to write

$$\langle v_{\text{esc}}^2 \rangle = \frac{8K}{M} = 4\langle v^2 \rangle, \tag{10.15}$$

where $\langle v^2 \rangle$ is the mass-weighted mean square velocity of the stars in the system. Thus, the root mean square escape speed is not an unattainably high multiple of the actual root mean square velocity of the stars; instead, it is only twice as fast. In particular, if two-body relaxation has given the stars a Maxwellian distribution of speeds (Equation 8.15), then the mean square velocity is $\langle v^2 \rangle = 3\sigma^2$. The fraction of stars moving faster than the escape speed, assuming a Maxwellian distribution, is then

$$F(v^2 > 12\sigma^2) \approx 0.00738 \approx \frac{1}{136}, \tag{10.16}$$

a number that Arthur Eddington would love.[1]

For a perfect Maxwellian distribution, this line of argument suggests that one out of 136 stars would rapidly leave the cluster, leaving a truncated Maxwellian. However, two-body relaxation replenishes the high-speed tail of the Maxwellian on a timescale $\sim t_{\text{rel,h}}$. The competition between slow replenishment and rapid depletion means that the loss rate of stars by evaporation is approximately

$$\frac{dN}{dt} \sim -\frac{1}{136} \frac{N}{t_{\text{rel,h}}}, \tag{10.17}$$

and the characteristic timescale for a system to lose most of its stars by evaporation is

$$t_{\text{evap}} \approx N \left| \frac{dN}{dt} \right|^{-1} \sim 136\, t_{\text{rel,h}}. \tag{10.18}$$

Thus, for a star cluster to evaporate in a time less than the Hubble time, it must have $t_{\text{rel,h}} < 100\,\text{Myr}$. Although some open clusters have half-mass relaxation times this short, they are also subject to the risk of being disrupted by molecular clouds, as discussed in Section 9.2. This means that few open clusters have the opportunity to evaporate quietly on their own.

10.2 Core Collapse

An intriguing end result of two-body relaxation is **core collapse**. Some globular clusters, such as NGC 6366 and 47 Tuc, have well-resolved central cores, as shown in Figure 8.2. When fitted with a King model, NGC 6366 has a core

---

[1] At least, he would have loved it in the year 1929, when he thought the fine structure constant was $\alpha = 1/136$; by 1930, he had transfered his affections to $1/137$.

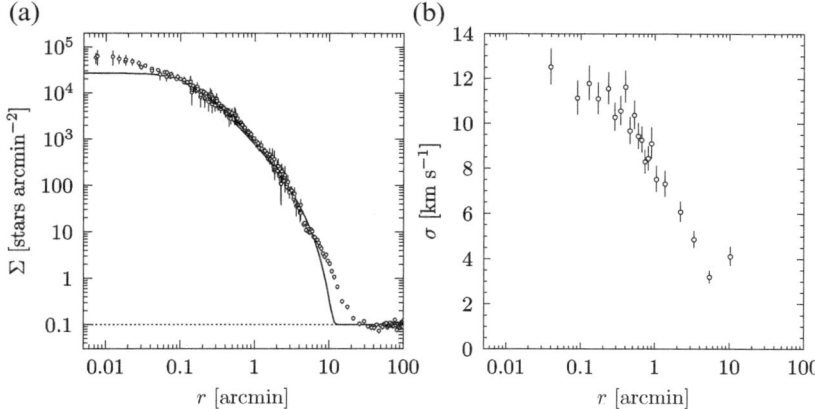

Figure 10.2 (a) Surface density of M15 (at $d \approx 10.5\,\text{kpc}$). The curved line is a King model with $W_0 = 8.2$ and $r_{\rm t} = 13.1\,\text{arcmin}$ (40 pc). [data from de Boer *et al.* 2019] (b) Observed line-of-sight velocity dispersion for M15. [data from Baumgardt and Hilker 2018]

radius $r_{\rm K} = 2.0\,\text{pc}$, while the more concentrated cluster 47 Tuc still has $r_{\rm K} = 0.5\,\text{pc}$. However, other globular clusters, such as the massive old cluster M15, lack an easily resolved core. As shown in Figure 10.2(a), although a King model fits the cluster's surface density reasonably well at intermediate radii, at smaller radii M15 lacks a large, well-defined core. Instead, the projected number density of stars continues to increase with decreasing radius, down to angular scales as small as $\sim 0.5\,\text{arcsec}$, corresponding to a physical size $r \sim 0.02\,\text{pc}$ at the distance of M15.

The lack of large, constant-density cores in some clusters is the result of a **gravothermal catastrophe** that causes core collapse. To see how two-body relaxation can lead to catastrophe, start with a globular cluster that has a central core with radius $r_{\rm c}$ and mass density $\rho_{\rm c}$. The core has a mass $M_{\rm c} \sim \rho_{\rm c} r_{\rm c}^3$, divided among $N_{\rm c}$ individual stars. Figure 10.2(b) shows the line-of-sight velocity dispersion for the cluster M15. Like most globular clusters, M15 has a velocity dispersion that decreases with radius; that is, its central regions are "hotter" than its outer regions.[2] Thus, we can think of a globular cluster as a spherical, self-gravitating object with a hot, dense core surrounded by a cooler, lower-density envelope. In this sense, it resembles an individual star, which is also a spherical, self-gravitating object with a hot, dense core surrounded by a cooler, lower-density envelope.

Within a star, energy must flow from the hot, dense core to the outer envelope. Similarly, within a globular cluster, energy must flow from the core to the envelope. However, in some obvious ways, a globular cluster cannot be

[2] Idealized King models are also hotter in their central regions, as shown in Figure 8.1(b).

treated as a gargantuan star. For instance, within a star, energy is transported by convection or radiation. Within a globular cluster, there are no convective instabilities that take low-dispersion clumps of stars into the core while taking high-dispersion clumps of stars out of the core. In addition, globular clusters are transparent (aside from the minuscule fraction of the volume occupied by opaque stars). Thus, there is no radiative transport of energy from the core of the globular cluster to its outer envelope. Thus, within a globular cluster, energy is transported by *conduction*. More precisely, a high-speed star from the center of the core will transfer energy to a lower-speed star by a gravitational encounter. This star will transfer energy to an even lower-speed star further out, and so forth. Since the transfer of energy is done by two-body gravitational encounters, the relevant timescale t_{flow} for energy to flow out of the core is the two-body relaxation time:

$$t_{\text{flow}} \sim t_{\text{rel,c}} \sim 0.12 \frac{N_c}{\ln(0.2N)} t_{\text{dyn,c}}, \qquad (10.19)$$

where $t_{\text{dyn,c}} = (G\rho_c)^{-1/2}$ is the dynamical time for the core.

Globular clusters are shaped like globes (that is, they are very nearly spherical); thus, we can apply Newton's shell theorems to them. In particular, we can treat the core of a globular cluster as a self-gravitating system; the outer spherical envelope has no net gravitational effect on the core. If the core is not rotating and is in virial equilibrium, the central velocity dispersion σ_c of the core can be equated to a temperature T_c by the relation

$$kT_c = m_\star \sigma_c^2, \qquad (10.20)$$

where $m_\star = M_c/N_c$ is the mean mass per star in the core. As we found in Section 6.1, a self-gravitating system has a *negative* heat capacity. Thus, as energy flows out of the core, its temperature T_c, or equivalently its velocity dispersion σ_c, must increase. If the surrounding envelope provides a large enough heat bath, the temperature difference between the core and envelope will increase. This increases the rate at which energy flows out of the core, and so triggers a runaway increase in the velocity dispersion σ_c of the core. This is the catastrophic result to which the term "gravothermal catastrophe" refers.

The evolution of a globular cluster can be calculated numerically. As early as 1980, Haldan Cohn simulated core collapse in a spherical isotropic cluster containing stars that all had the same mass. The results of Cohn's calculation are shown in Figure 10.3. Regardless of the exact initial conditions, the density profile of the cluster evolves into a **self-similar** solution. That is, at any time t, the density profile has the same shape; it simply moves up and to the left in Figure 10.3 as time goes on. The central density $\rho_c(t)$ increases as core collapse proceeds, while the core radius $r_c(t)$ decreases. The self-similar solution for the density profile can be written in the form

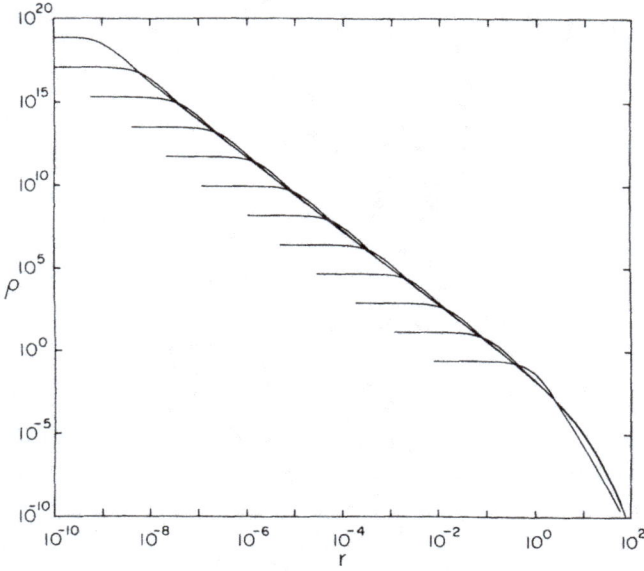

Figure 10.3 Evolution of the density profile of a spherical isotropic globular cluster. As time increases, the core radius decreases and the central density increases. [Cohn 1980]

$$\rho(r, t) = \rho_c(t) f_\rho(x), \tag{10.21}$$

where f_ρ is a dimensionless function of the dimensionless radius $x \equiv r/r_c(t)$. The numerical results show that outside the core, $\rho(r) \propto r^{-\beta}$, where $\beta \approx 2.23$. Thus,

$$f_\rho(x) \propto x^{-\beta} \qquad [x \gg 1] \tag{10.22}$$

while the central density is related to the core radius by

$$\rho_c(t) \propto r_c(t)^{-\beta}. \tag{10.23}$$

The mass of the core then obeys the relation

$$M_c \propto \rho_c r_c^3 \propto r_c^{3-\beta} \propto r_c^{0.77}, \tag{10.24}$$

and the core mass decreases as the core radius shrinks.

The time τ required to collapse from an initial core radius r_c to an infinitesimal core radius is proportional to the two-body relaxation time of the core, since the energy flow is driven by two-body gravitational encounters. Therefore, if we ignore the logarithmic factor in Equation 10.19, we can make the approximation

$$\tau \sim 0.12 N_c (G\rho_c)^{-1/2} \propto M_c \rho_c^{-1/2} \propto r_c^{3-\beta/2} \propto r_c^{1.88}. \tag{10.25}$$

Inverting Equation 10.25 to find the core radius in terms of the time τ until core collapse is complete, we find

$$r_c \propto \tau^{2/(6-\beta)} \propto \tau^{0.53}. \tag{10.26}$$

Therefore, as the clock ticks down to $\tau = 0$, the central density, core mass, and central dispersion evolve as

$$\rho_c \propto r_c^{-\beta} \propto \tau^{-2\beta/(6-\beta)} \propto \tau^{-1.18}, \tag{10.27}$$

$$M_c \propto \rho_c r_c^3 \propto \tau^{2(3-\beta)/(6-\beta)} \propto \tau^{0.41}, \tag{10.28}$$

$$\sigma_c \propto (M_c/r_c)^{1/2} \propto \tau^{(2-\beta)/(6-\beta)} \propto \tau^{-0.06}. \tag{10.29}$$

Although the velocity dispersion in the core does increase as core collapse progresses, there is not a strong dependence on τ. The more dramatic expression of core collapse is the steep increase in the central density of the globular cluster.

As catastrophes go, the gravothermal catastrophe is disappointingly tame; in the words of Binney and Tremaine, core collapse is "a relatively unspectacular process." Formally, the self-similar solution of Cohn leads to the formation of a black hole; since the Schwarzschild radius of the core is $R_{\text{sch}} \propto M_c$, we find that

$$\frac{r_c}{R_{\text{sch}}} \propto \frac{r_c}{M_c} \propto \tau^{-2(2-\beta)/(6-\beta)} \propto \tau^{0.12}. \tag{10.30}$$

However, consider a real globular cluster such as 47 Tucanae, which has a core radius $r_c \approx 0.5$ pc, a central velocity dispersion $\sigma_c \approx 12 \, \text{km s}^{-1}$, and a core mass $M_c \sim 3\sigma_c^2 r_c/G \sim 50\,000 \, M_\odot$. Shrinking this core mass down to the mass of a single star, $M_c \sim 0.5 \, M_\odot$, requires decreasing the core mass by a factor of $\sim 10^{-5}$. The self-similar solution states that the core radius also shrinks, but by a factor of $\sim 10^{-5/(3-\beta)} \sim 3 \times 10^{-7}$. The core radius thus decreases from $r_c \approx 0.5$ pc to a much more petite $r_c \sim 0.03$ au. However, even this relatively small core size is vastly larger than the Schwarzschild radius of a star ($R_{\text{sch}} \sim 10^{-8}$ au). Therefore, the assumptions of stellar dynamics will break down long before the core shrinks to its Schwarzschild radius.

One puzzle must be addressed when we examine core collapse: many globular clusters have sizeable cores despite being much older than the relaxation time for their core. As an example, let's take another look at the cluster 47 Tuc. Its dense core, with $M_c \sim 50\,000 \, M_\odot$ of stars within a radius $r_c \approx 0.5$ pc, has a relatively short dynamical time, with $t_{\text{dyn,c}} \sim r_c/\sigma_c \sim 0.05$ Myr. Of the $N \sim 2 \times 10^6$ stars in 47 Tuc, a fraction $f \sim 0.06$ are in the core. This yields a core relaxation time of $t_{\text{rel,c}} \sim 1000 t_{\text{dyn,c}} \sim 50$ Myr. Other globular clusters with sizeable cores, such as NGC 6366 (Figure 8.2(a)), also have core relaxation times much shorter than their age. What prevents core collapse in these systems from reaching its end state? Let's return to the analogy between a cluster of stars and an individual star. They are both self-gravitating spherical systems with a dense central core and a more tenuous outer envelope. They are both subject to a gravothermal instability that can lead to core collapse. In a star, however, the process of core collapse is paused for as long as nuclear fusion provides an energy source at the star's center, replacing the energy that it carried outward by convection and radiation. Similarly, core collapse in a globular cluster can be paused if there is an energy source in the cluster's core.

In practice, the energy source that pauses core collapse in globular clusters consists of a population of binary star systems. A single close binary, with stars of equal mass m and semimajor axis a, has an energy (Equation 1.54)

$$E_{\text{orb}} = -\frac{Gm^2}{2a} = -4.7 \times 10^{47} \, \text{erg} \left(\frac{m}{0.5 \, \text{M}_\odot}\right)^2 \left(\frac{a}{1 \, \text{R}_\odot}\right)^{-1}. \qquad (10.31)$$

We can compare this energy to the total energy of a large globular cluster like 47 Tuc. Using the virial theorem, the cluster energy is $E = -W/2$, or

$$E \approx -\frac{GM^2}{4r_{\text{h}}} \approx -4.3 \times 10^{51} \, \text{erg} \left(\frac{M}{10^6 \, \text{M}_\odot}\right)^2 \left(\frac{r_{\text{h}}}{5 \, \text{pc}}\right)^{-1}. \qquad (10.32)$$

Thus, taking $\sim 10^4$ wide binaries and shrinking them until they are contact binaries will release enough energy (if it can be harnessed) to unbind a large globular cluster.

To examine how energy can be removed from, or added to, a binary star system, take our sample binary, with stellar mass m and separation a, and place it in a larger stellar system. In the binary's neighborhood, the mean stellar mass is $m_\star$ and the velocity dispersion is σ. The ultimate fate of the binary depends on whether it is **hard** or **soft**. The division between hard and soft depends on the properties of the surrounding stellar system. The standard division is

$$|E_{\text{orb}}| > m_\star \sigma^2 \qquad \text{[hard]}, \qquad (10.33)$$

$$|E_{\text{orb}}| < m_\star \sigma^2 \qquad \text{[soft]}. \qquad (10.34)$$

If $m \approx m_\star$, then the critical orbit size dividing hard (close) binaries from soft (distant) binaries is

$$a \approx \frac{Gm_\star}{2\sigma^2} \approx 8.9 \, \text{au} \left(\frac{m_\star}{0.5 \, \text{M}_\odot}\right) \left(\frac{\sigma}{5 \, \text{km s}^{-1}}\right)^{-2}, \qquad (10.35)$$

scaling to a dispersion typical of globular clusters.

In 1975, Douglas Heggie and Jack Hills separately studied the problem of the evolution of binaries embedded within a larger stellar system. Their findings can be summed up by the Heggie–Hills law: "The soft get softer and the hard get harder." To look at the softening and hardening process in more detail, consider the interaction between our sample binary and an individual field star. One possible outcome is the complete disruption of the binary system by the field star; this outcome is metaphorically called the **ionization** of the binary. If a and b are the two members of the binary, and c is the field star, the ionization can be written symbolically as

$$ab + c \rightarrow a + b + c. \tag{10.36}$$

Energetically, this outcome requires that $m_\star v^2/2 > |E_{orb}|$, where v is the field star's initial speed relative to the barycenter of the binary. Thus, ionization is only a likely outcome for soft binaries. Calculations of the mean time until ionization, t_{ion}, reveal that, for very soft binaries,

$$\frac{t_{ion}}{t_{rel}} \approx 3 \frac{|E_{orb}|}{m_\star \sigma^2} \qquad [|E_{orb}| \ll m_\star \sigma^2]. \tag{10.37}$$

In the opposite limit of very hard binaries,

$$\frac{t_{ion}}{t_{rel}} \approx 10 \left(\frac{E_{orb}}{m_\star \sigma^2} \right)^2 \exp \left(\frac{3}{4} \frac{|E_{orb}|}{m_\star \sigma^2} \right) \qquad [|E_{orb}| \gg m_\star \sigma^2]. \tag{10.38}$$

Since ionization of a hard binary requires a field star on the exponential high-speed tail of a Maxwellian distribution, the binary's lifetime before ionization is exponentially longer than the local two-body relaxation time.

    An ionization is a less likely outcome than a distant **flyby** that leaves the binary system intact:

$$ab + c \rightarrow ab + c. \tag{10.39}$$

On average, a hard binary will lose energy from such an encounter, while a soft binary will gain energy. Numerous flybys will gradually increase the energy of the soft binary until it becomes unbound; this outcome is metaphorically called the **evaporation** of the binary. The time required for evaporation of a soft binary is

$$\frac{t_{evap}}{t_{rel}} \approx 0.3 \frac{|E_{orb}|}{m_\star \sigma^2} \qquad [|E_{orb}| < m_\star \sigma^2]. \tag{10.40}$$

This timescale is shorter by a factor of $\sim 1/10$ than the expected time before ionization by a single encounter (Equation 10.37). Thus, soft binaries will most likely be gradually nudged apart by numerous distant encounters, rather than being ripped rudely asunder by a single close encounter. Since the evaporation time for soft binaries is shorter than the two-body relaxation time, we expect an absence of soft binaries in collisional systems (unless, of course, there is a mechanism for creating soft binaries).

    A final possible outcome of an interaction between a field star and a binary is an **exchange**:

$$ab + c \rightarrow a + bc. \tag{10.41}$$

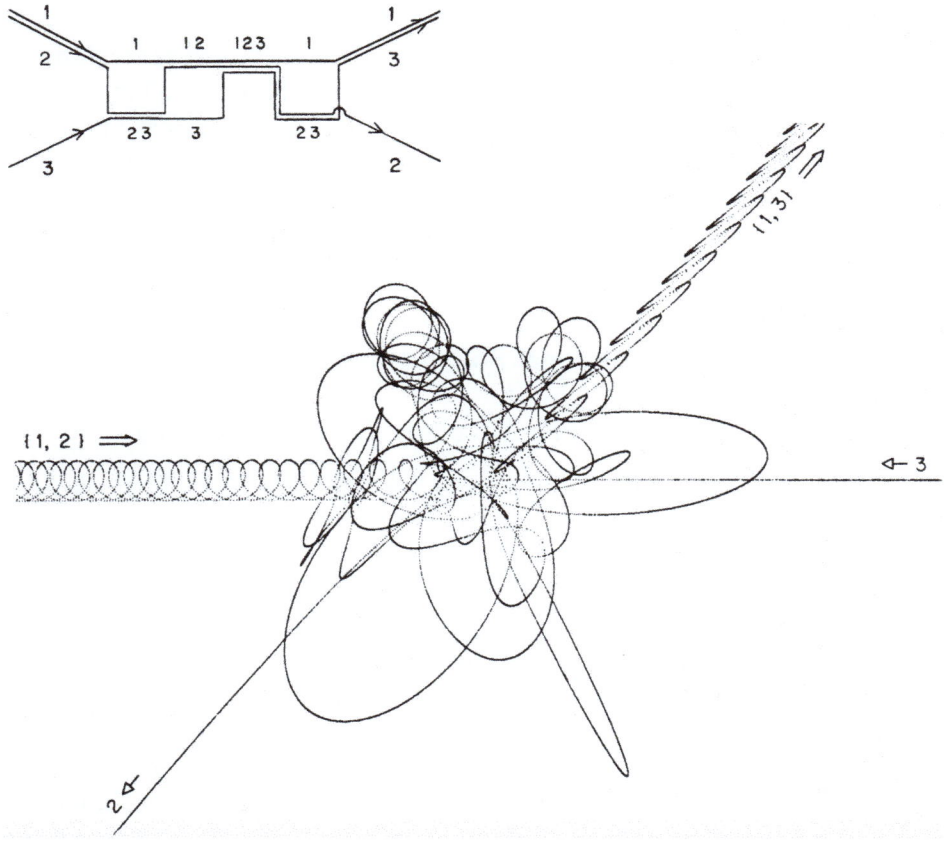

Figure 10.4 Example of a resonant exchange between a binary and a field star. The dots along the trajectories represent equal time intervals. [Hut and Bahcall 1983]

An exchange can be either prompt, taking place over a time comparable to the orbital period of the binary, or resonant. Resonant exchanges often take the form of an intricate but temporary three-body dance. Figure 10.4 depicts a particularly simple resonant exchange, in which all three stars have the same mass, and all stellar motion lies in a single plane. More generally, a three-dimensional resonant exchange can be far more tangled in appearance as the chaotic dance plays out.

Adding all three types of interactions – ionization, flyby, and exchange – the average rate of change of a binary's energy is

$$\frac{dE_{\rm orb}}{dt} \approx -0.2\frac{m_\star \sigma^2}{t_{\rm rel}} \qquad \text{[hard]}, \tag{10.42}$$

$$\frac{dE_{\rm orb}}{dt} \approx +3\frac{m_\star \sigma^2}{t_{\rm rel}} \qquad \text{[soft]}. \tag{10.43}$$

Hard binaries thus act as a heat source; as hard binaries get harder, they transfer energy to the random kinetic energy of field stars. The central regions of globular clusters are particularly rich in hard binaries. Since hard binaries are firmly held together by gravity, they act as "double-mass" stars as far as the mass segregation process is concerned. Once again, the analogy between a star cluster and an individual star is useful: just as the fusion energy source in a star is concentrated toward its center, the hard binary energy source in a globular cluster is also concentrated toward the center.

10.3 Collisional Boltzmann Equation

In Section 6.2, it was asserted that the collisionless Boltzmann equation,

$$\frac{\partial f}{\partial t} + \vec{v} \cdot \vec{\nabla} f - \vec{\nabla} \Phi \cdot \frac{\partial f}{\partial \vec{v}} = 0, \tag{10.44}$$

is a very important equation for stellar dynamics – and so it is. However, the collisionless Boltzmann equation has one intrinsic shortcoming: it deals only with collisionless systems.

Stellar systems that are older than their two-body relaxation time cannot be treated as purely collisionless systems. In addition to open clusters and globular clusters, these collisional systems include the dense nuclei of galaxies. If we want the Boltzmann equation to describe processes such as core collapse, cluster evaporation, and mass segregation, we must add a term to the equation that includes the effects of two-body interactions. In general, the **collisional Boltzmann equation** is written as

$$\frac{\partial f}{\partial t} + \vec{v} \cdot \vec{\nabla} f - \vec{\nabla} \Phi \cdot \frac{\partial f}{\partial \vec{v}} = \Gamma(f), \tag{10.45}$$

where $\Gamma(f)$ is the **collision term**. For a complete understanding of collisional stellar systems, we need to find the appropriate functional form for the collision term. This is not an easy job.

To find what form $\Gamma(f)$ takes, start by contemplating the physics of a system in which gravitational encounters between individual stars play an important role. Since encounters between stars kick them in random directions, we can think of the net result of gravitational encounters as being a **random walk** through phase space. The random walk of stars through phase space has a useful analogy with the diffusion of molecules through the air as a result of collisions with other molecules. The usefulness of this analogy results primarily from the fact that we can appropriate the mathematical tools that have been developed to describe diffusion.

To keep the notation simple, let $\vec{w} \equiv (\vec{r}, \vec{v})$ be the six-dimensional vector that specifies the position of a star in phase space. We also define a **transition probability** Ψ_t such that

$$\Psi_t(\vec{w}, \Delta\vec{w})d^6(\Delta\vec{w}) \qquad (10.46)$$

is the probability per unit time that a star at phase space location $\vec{w}$ will be kicked into a volume element $d^6(\Delta\vec{w})$ at a phase space location $\vec{w} + \Delta\vec{w}$. Note that $\Delta\vec{w}$ doesn't have to be small in amplitude; a close gravitational encounter can give a big velocity kick, for instance. Stars will be scattered away from the phase space location $\vec{w}$ at a rate

$$\left.\frac{\partial f}{\partial t}\right|_- = -\int \Psi_t(\vec{w}, \Delta\vec{w})f(\vec{w})\, d^6(\Delta\vec{w}). \qquad (10.47)$$

Stars will be scattered into the phase space location $\vec{w}$ at a rate

$$\left.\frac{\partial f}{\partial t}\right|_+ = \int \Psi_t(\vec{w} - \Delta\vec{w}, \Delta\vec{w})f(\vec{w} - \Delta\vec{w})\, d^6(\Delta\vec{w}). \qquad (10.48)$$

Thus, the net collision term will be

$$\Gamma(f) = \left.\frac{\partial f}{\partial t}\right|_- + \left.\frac{\partial f}{\partial t}\right|_+. \qquad (10.49)$$

Unfortunately, Equation 10.49 is one of those equations that are so general as to be useless in practice. The transition probability Ψ_t results from a mix of close and distant gravitational encounters with stars that have not necessarily relaxed to a Maxwellian velocity distribution. To make progress, we need some simplifying assumptions about the physics of the problem.

Our first assumption will be that the collision term $\Gamma(f)$ results primarily from distant, weak gravitational encounters; that is, encounters with $b \gg b_{\text{cge}} \sim Gm_\star/v^2$. This assumption is known as the **Fokker–Planck approximation**.[3] For a weak encounter, we can assume that $\Delta\vec{w}$ is small, and do a Taylor series expansion of the quantity

$$\Psi_t(\vec{w} - \Delta\vec{w}, \Delta\vec{w})f(\vec{w} - \Delta\vec{w}) \qquad (10.50)$$

in Equation 10.48. If the expansion is taken as far as terms of order $(\Delta w)^2$, we find the **Fokker–Planck equation**:

$$\Gamma(f) = -\sum_i \frac{\partial}{\partial w_i}[A(\Delta w_i)f(\vec{w})] + \frac{1}{2}\sum_{i,j} \frac{\partial^2}{\partial w_i \partial w_j}[D(\Delta w_i \Delta w_j)f(\vec{w})]. \qquad (10.51)$$

The terms labeled $A(\Delta w_i)$ in Equation 10.51 are **drift coefficients** that describe the steady systematic drift in the six dimensions of phase space. The terms labeled $D(\Delta w_i \Delta w_j)$ are **diffusion coefficients** that describe the random walk of a star through phase space. Since $D(\Delta w_i \Delta w_j) = D(\Delta w_j \Delta w_i)$, the diffusion coefficients form a symmetric 6×6 matrix with 21 independent components.

[3] The "Fokker" of the Fokker–Planck approximations is the Dutch musician/physicist Adriaan Fokker, not to be confused with his cousin Anthony, the aeronautical engineer who designed the Fokker Dr.1 tri-plane.

In many cases, we can reduce the number of drift and diffusion coefficients by adopting the **local approximation**, which states that a test star's drift and diffusion result primarily from local encounters with $b \ll b_{\max} \sim R$, where R is the size of the stellar system in question. (Remember, we have already embraced the Fokker–Planck approximation, which states that $b \gg b_{\mathrm{cge}}$.) If the local approximation holds true, then encounters take place on a timescale $t \sim b/\sigma$ that is short compared to the dynamical time $t \sim R/\sigma$ for the stellar system as a whole. In this case, we can assume that all gravitational encounters are impulsive, producing an instantaneous change $\Delta\vec{v}$ in a star's velocity, but not an instantaneous change $\Delta\vec{r}$ in its position. Thus, the local approximation leads to

$$A(\Delta r_i) = D(\Delta r_i \Delta x_j) = D(\Delta r_i \Delta v_j) = 0. \qquad (10.52)$$

This is a helpful simplification, leaving us with only three non-zero drift coefficients, and six non-zero diffusion coefficients.

The picture becomes simpler still if the test star moves through a stellar system with an isotropic velocity dispersion. In that case, the only non-zero drift coefficient is $A(\Delta v_{\parallel})$, where $\Delta v_{\parallel}$ is the component of the velocity kick parallel to the star's direction of motion. On average, however, the velocity kicks $\Delta v_{\perp}$ from all stellar encounters will cancel out; for every kick to the right, there will be a kick to the left (on average). In fact, we are using the same physics arguments we made in Section 9.1 when discussing the slowing of a compact object by dynamical friction as it moved through a field of stars, each with mass $m_{\star}$. The only difference is that now we are examining an object with a mass $m \sim m_{\star}$, instead of a mass $M \gg m_{\star}$. If the field stars have a Maxwellian velocity distribution, with one-dimensional velocity dispersion σ, then the drift coefficient $A(\Delta v_{\parallel})$ can be computed analytically. It becomes

$$A(\Delta v_{\parallel}) = -C_0 \frac{1 + m/m_{\star}}{\sqrt{2}\sigma} G_{\mathrm{df}}(X), \qquad (10.53)$$

where the normalization constant is

$$C_0 = 4\sqrt{2\pi} \frac{G^2 \rho_{\star} m_{\star} \ln \Lambda}{\sigma}, \qquad (10.54)$$

the dimensionless speed of the test star is $X = v/(\sqrt{2}\sigma)$, and $G_{\mathrm{df}}(X)$ is the same dynamical friction function that we encountered in Equation 9.19:

$$G_{\mathrm{df}}(X) = \frac{1}{2X^2} \left[\mathrm{erf}(X) - \frac{2}{\sqrt{\pi}} X \exp(-X^2) \right]. \qquad (10.55)$$

Since the test star is undergoing dynamical friction, the drift coefficient in Equation 10.53 is negative at any speed. The absolute value of $A(\Delta v_{\parallel})$ is shown in Figure 10.5(a); note that it exactly matches the shape of the dynamical friction deceleration as plotted in Figure 9.3.

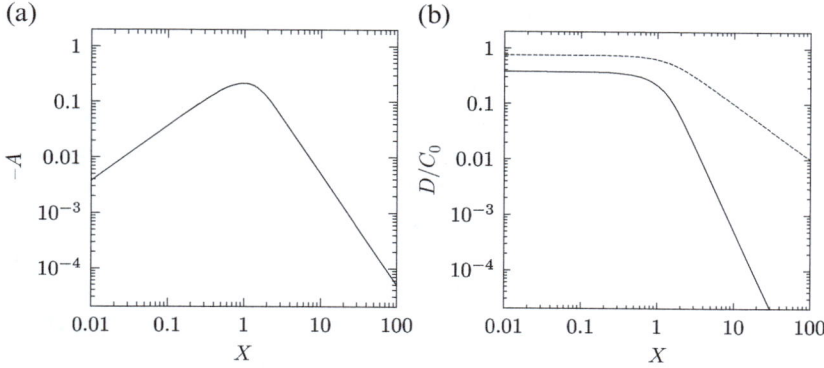

Figure 10.5 (a) Drift coefficient $-A(\Delta v_\parallel)$, in arbitrary units, for a Maxwellian velocity distribution. (b) Diffusion coefficient $D(|\Delta \vec{v}_\perp|^2)$, shown as the dashed line, and $D(\Delta v_\parallel^2)$, shown as the solid line. The diffusion coefficients are in units of the constant C_0.

If we still assume an isotropic velocity dispersion for the field stars, the only preferred direction is that of the test star's motion. This means that the only non-zero diffusion coefficients are $D(\Delta v_\parallel^2)$ and $D(|\Delta \vec{v}_\perp|^2)$; here, $\vec{v}_\perp$ is the velocity kick perpendicular to the test star's direction of motion. If the field stars have a Maxwellian velocity distribution, then the values of the diffusion coefficients are

$$D(|\Delta \vec{v}_\perp|^2) = C_0 \left(\frac{\mathrm{erf}(X) - G_{\mathrm{df}}(X)}{X} \right) \tag{10.56}$$

and

$$D(\Delta v_\parallel^2) = C_0 \left(\frac{G_{\mathrm{df}}(X)}{X} \right). \tag{10.57}$$

The diffusion coefficient in the perpendicular direction (Equation 10.56) is shown as the dashed line in Figure 10.5(b); for high-speed test stars, with $X \gg 1$, it falls off as $D \propto X^{-1}$. The diffusion coefficient in the parallel direction (Equation 10.57) is shown as the solid line in Figure 10.5(b); at $X \gg 1$, it falls steeply, with $D \propto X^{-3}$. At any speed, diffusion in the direction perpendicular to $\vec{v}$ is more effective than in the direction parallel to $\vec{v}$.

The Fokker–Planck equation can be used to estimate the rate at which a test star's kinetic energy, $K = mv^2/2$, changes with time as the result of gravitational encounters with other stars. If the field stars have an isotropic velocity dispersion, then

$$\frac{1}{m} \frac{dK}{dt} = \frac{1}{2} D(\Delta v_\parallel^2) + \frac{1}{2} D(|\Delta \vec{v}_\perp|^2) + v A(\Delta v_\parallel). \tag{10.58}$$

If the field stars have a Maxwellian distribution, this becomes

$$\frac{1}{m}\frac{dK}{dt} = C_0 \left[\frac{1 + m/m_\star}{\sqrt{\pi}} \exp(-X^2) - \frac{m}{m_\star}\frac{\mathrm{erf}(X)}{2X} \right]. \qquad (10.59)$$

A slow test star, with $X \ll 1$, will tend to gain kinetic energy as it randomly walks through velocity space; its rate of change in kinetic energy is

$$\frac{1}{m}\frac{dK}{dt} \approx \frac{C_0}{\sqrt{\pi}} \qquad [X \ll 1]. \qquad (10.60)$$

By contrast, a fast test star, with $X \gg 1$, will tend to lose kinetic energy by dynamical friction; its rate of change in kinetic energy is

$$\frac{1}{m}\frac{dK}{dt} \approx -\frac{C_0}{2}\frac{m}{m_\star}\frac{1}{X} \qquad [X \gg 1]. \qquad (10.61)$$

Figure 10.6 shows the rate of change in kinetic energy for all values of X, assuming a test star equal in mass to the field stars ($m = m_\star$). A modified Robin Hood effect is at work. Dynamical friction takes energy most effectively from the "middle class" stars with $v^2/2 \sim 2\sigma^2$, while diffusion in velocity space gives energy preferentially to the "poor" stars with $v^2/2 < \sigma^2$. Although the "rich" stars with $v^2/2 \gg \sigma^2$ do lose energy by dynamical friction, the timescale for them to lose all their kinetic energy is

$$t_{\mathrm{df}} = K \left| \frac{dK}{dt} \right|^{-1} \propto v^{3/2}, \qquad (10.62)$$

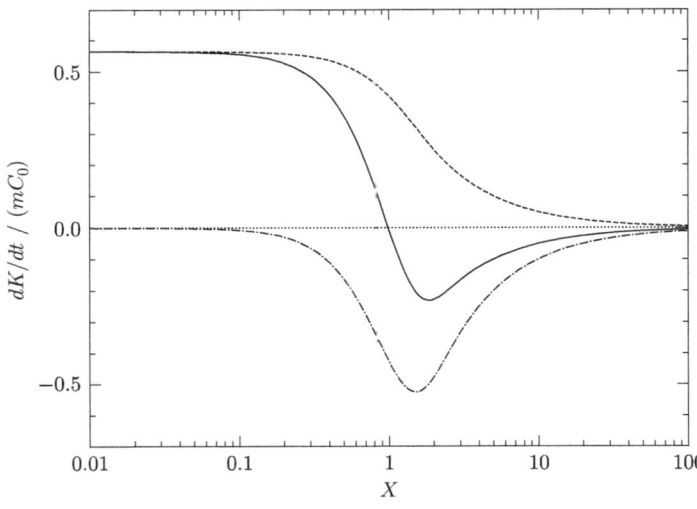

Figure 10.6 Rate at which a star loses energy from dynamical friction (dot-dashed line), rate at which it gains energy from diffusion in phase space (dashed line), and net rate of change in its kinetic energy (solid line).

making use of Equation 10.61. Although the net effect of two-body interactions is to drive stellar velocities to the equilibrium point at $X = 1$, it takes a long time for a star in the high-speed exponential tail of the Maxwellian to be slowed by dynamical friction; so long, in fact, that another star will have won the random walk lottery and replaced it in the high-speed tail.

Viewed from a great distance, a relaxed stellar system looks placid and unchanging. However, for any individual star there is always the possibility of a gravitational encounter to change its fortune dramatically.

Exercises

10.1 To compute the Kozai cycles illustrated in Figure 2.10, we impudently gave the Sun a twin companion star with $m_2 = 1\,M_\odot$.
  (a) Scaling to a one-dimensional velocity dispersion $\sigma_\star = 30\,\mathrm{km\,s^{-1}}$ for stars in the solar neighborhood, at what separation a would a binary consisting of the Sun and its twin switch from "hard" to "soft"?
  (b) Scaling to a number density $n_\star = 0.2\,\mathrm{pc^{-3}}$ for stars in the solar neighborhood, at what separation a would the evaporation time t_{evap} for the Sun and its twin be equal to the current age of the Sun, $t = 4.57\,\mathrm{Gyr}$?

10.2 You have discovered the secret of immortality, and have decided to spend your elongated life dismantling the globular cluster 47 Tucanae. Before you start your task, 47 Tuc has a mass $M = 10^6\,M_\odot$ and half-mass radius $r_{\mathrm{h}} = 5\,\mathrm{pc}$. Once per century thereafter, you identify the most weakly bound star in the cluster, and accelerate it to the escape speed from its location.
  (a) Assume every star in the cluster has a mass $m_\star = 0.5\,M_\odot$. Is the dismantling time, $t_{\mathrm{dism}} = M/\dot{M}$, longer than or shorter than the initial dynamical time at the half-mass radius? Is the dismantling time longer than or shorter than the initial two-body relaxation time at the half-mass radius?
  (b) The first few thousand stars that you remove are very weakly bound; thus, removing them doesn't significantly change the total energy $E = K + W$ of the cluster. With the approximations $E = \mathrm{constant}$ and $\dot{M} = \mathrm{constant}$, how does the half-mass radius $r_{\mathrm{h}}(t)$ evolve with time? Explicitly state any other approximations that you make.

Appendix

Constants and Units

Table A.1 Fundamental physical constants.

Name	Symbol	Value	Units
Speed of light in vacuum	c	$2.997\,924\,58 \times 10^{10}$	cm s^{-1}
Gravitation constant	G	$6.674\,30 \times 10^{-8}$	$\text{cm}^3\,\text{g}^{-1}\,\text{s}^{-2}$
Planck constant	h	$6.626\,070\,15 \times 10^{-27}$	erg s
		$4.135\,667\,696 \times 10^{-15}$	eV s
Reduced Planck constant	$\hbar$	$1.054\,571\,817 \times 10^{-27}$	erg s
		$6.582\,119\,569 \times 10^{-16}$	eV s
Boltzmann constant	k	$1.380\,649 \times 10^{-16}$	erg K^{-1}
		$8.617\,333\,262 \times 10^{-5}$	eV K^{-1}
Elementary charge	e	$1.602\,176\,634 \times 10^{-19}$	C
Stefan–Boltzmann constant	σ_{SB}	$5.670\,374\,419 \times 10^{-5}$	$\text{erg cm}^{-2}\,\text{s}^{-1}\,\text{K}^{-4}$
Thomson cross section	σ_e	$6.652\,458\,705\,1 \times 10^{-25}$	cm^2
Atomic mass unit	m_{AMU}	$1.660\,539\,068\,92 \times 10^{-24}$	g
Proton mass	m_p	$1.672\,621\,925\,95 \times 10^{-24}$	g
Electron mass	m_e	$9.109\,383\,713\,9 \times 10^{-28}$	g
Proton magnetic moment	μ_p	$1.401\,606\,795\,45 \times 10^{-23}$	erg G^{-1}
Electron magnetic moment	μ_e	$-9.284\,764\,691\,7 \times 10^{-21}$	erg G^{-1}
Bohr radius	a_0	$5.291\,772\,105\,44 \times 10^{-9}$	cm

Source: 2022 CODATA recommended values.

Table A.2 Astronomical constants.

Name	Symbol	Value	Units
Astronomical unit	au	$1.495\,978\,707 \times 10^{13}$	cm
Solar radius	$R_\odot$	6.957×10^{10}	cm
Solar luminosity	$L_\odot$	3.828×10^{33}	erg s^{-1}
Solar gravitational parameter	$GM_\odot$	$1.327\,124\,4 \times 10^{26}$	$\text{cm}^3\,\text{s}^{-2}$
Solar mass	$M_\odot$	1.9884×10^{33}	g
Solar effective temperature	$T_\odot$	5772	K

Sources: IAU 2012 resolution B2 on the redefinition of the astronomical unit of length; IAU 2015 resolution B3 on recommended nominal conversion constants for solar and planetary properties.

Table A.3 Astronomical and physical units.

Name	Symbol	Value	Units
Parsec	pc	206 264.8063	au
		$3.085\,677\,58 \times 10^{18}$	cm
Day	d	86 400	s
Julian year	yr	365.25	d
		$3.155\,76 \times 10^{7}$	s
Angstrom	Å	10^{-8}	cm
Electron volt	eV	$1.602\,176\,634 \times 10^{-12}$	erg
Standard atmosphere	atm	$1.013\,25 \times 10^{6}$	$\mathrm{dyn\,cm^{-2}}$

Sources: IAU 2012 resolution B2; IAU 2015 resolution B3; 2022 CODATA recommended values.

Further Reading, Bibliography, and Figure Credits

Merre ocur Pangur bán,	*Pangur Ban the cat and I*
Cectar nator fria rainán;	*Pursue our craft without a pause.*
b´c a menma-ram fri reilgg,	*My fingers on the keyboard fly;*
mu menma céin im raincerroo.	*He hunts the mouse with flashing claws.*
. . .	. . .
hé ferin ar coimrio dáu	*On my laptop day and night,*
In muio ou-ngn´ cac óenláu;	*Hunting bugs is what he knows,*
Ou caibairc dorairo ou glé	*While I find my own delight*
for mu muio céin am merre.	*Compiling my dynamic prose.*

    – Ninth-century Irish poem   – Loose modern adaptation

Further Reading

Binney, J., and Tremaine, S., 2008, *Galactic Dynamics* (2nd edition), Princeton University Press

Murray, C. D., and Dermott, S. F. 1999, *Solar System Dynamics*, Cambridge University Press

Tremaine, S., 2023, *Dynamics of Planetary Systems*, Princeton University Press

Bibliography

In our reference list, we use compact abbreviations for the most frequently cited journals in astronomy. These abbreviations, as used by the SAO/NASA Astrophysics Data System, are:

- A&A = Astronomy and Astrophysics
- AJ = The Astronomical Journal
- ApJ = The Astrophysical Journal
- ApJS = The Astrophysical Journal Supplement
- ARA&A = Annual Review of Astronomy and Astrophysics
- MNRAS = Monthly Notices of the Royal Astronomical Society
- PASP = Publications of the Astronomical Society of the Pacific

Baumgardt, H., and Hilker, M. 2018, "A catalogue of masses, structural parameters, and velocity dispersion profiles of 112 Milky Way globular clusters," MNRAS, 478, 1520

Baumgardt, H., Hilker, M., Sollima, A., *et al.* 2019, "Mean proper motions, space orbits, and velocity dispersion profiles of Galactic globular clusters derived from Gaia DR2 data," MNRAS, 482, 5138

Birkby, J. L., de Kok, R. J., Brogi, M., *et al.* 2017, "Discovery of water at high spectral resolution in the atmosphere of 51 Peg," AJ, 153, 138

Boltzmann, L. 1872, "Weitere Studien über das Wärmegleichgewicht unter Gasmolekülen," Sitzungsberichte Akad. Wiss., 66, 275

Bonaca, A., Pearson, S., Price-Whelan, A. M., *et al.* 2020, "Variations in the width, density, and direction of the Palomar 5 tidal tails," ApJ, 889, 70

Bovy, J. 2015, "galpy: A python library for galactic dynamics," ApJS, 216, 29

Chandrasekhar, S. 1943, "Dynamical friction. I. General considerations: The coefficient of dynamical friction," ApJ, 97, 255

Clausius, R. 1870, "On a mechanical theorem applicable to heat," Philosophical Magazine, fourth series, 40, 122 [translated from a presentation before the Niederrheinischen Gesellschaft für Natur- und Heilkunde, 1870 Jun 13]

Cohn, H. 1980, "Late core collapse in star clusters and the gravothermal instability," ApJ, 242, 765

Dawson, R. I., Johnson, J. A., Fabrycky, D. C., *et al.* 2014, "Large eccentricity, low mutual inclination: The three-dimensional architecture of a hierarchical system of giant planets," ApJ, 791, 89

de Boer, T. J. L., Gieles, M., Balbinot, E., *et al.* 2019, "Globular cluster number density profiles using *Gaia* DR2," MNRAS, 485, 4906

Eddington, A. S. 1916, "The kinetic energy of a star cluster," MNRAS, 76, 525

Harris, W. E. 1996, "A catalog of parameters for globular clusters in the Milky Way," AJ, 112, 1487

Heggie, D. C. 1975, "Binary evolution in stellar dynamics," MNRAS 173, 729

Hills, J. G. 1975, "Encounters between binary and single stars and their effect on the dynamical evolution of stellar systems," AJ, 80, 809

Holczer, T., Mazer, T., Nachmani, G., *et al.* 2016, "Transit timing observations from Kepler. IX. Catalog of the full long-cadence data set," ApJS, 225, 9

Hut, P., and Bahcall, J. N. 1983, "Binary-single star scattering. I. Numerical experiments for equal masses," ApJ, 268, 319

Jeans, J. H. 1902, "The stability of a spherical nebula," Phil. Trans. R. Soc., 199, 1

Jeans, J. H. 1915, "On the theory of star-streaming and the structure of the universe," MNRAS, 76, 70

King, I. R. 1966, "The structure of star clusters. III. Some simple dynamical models," ApJ, 71, 64

Kozai, Y. 1962, "Secular perturbations of asteroids with high inclination," AJ, 31, 185

Lidov, M. L. 1962, "The evolution of orbits of artificial satellites of planets under the action of gravitational perturbations of external bodies," Planetary & Space Science, 9, 719 [translated from Iskusstvennye Sputniki Zemli, 1961, 8, 5]

Lin, C. C., and Shu, F. 1964, "On the spiral structure of disk galaxies," ApJ, 140, 646

Lissajous, J. 1857, "Mémoire sur l'étude optique des mouvements vibratoires," Annales de Chimie et de Physique (series 3), 51, 147

Martinazzi, E., Pieres, A., Kepler, S. O., *et al.* 2014, "Probing mass segregation in the globular cluster NGC 6397," MNRAS, 442, 3105

Mestel, L. 1963, "On the galactic law of rotation," MNRAS, 126, 553

Moutou, C., Mayor, M., Lo Curto, G., *et al.* 2009, "The HARPS search for southern extra-solar planets. XV.," A&A, 496, 513

Navarro, J. F., Frenk, C. S., and White, S. D. M. 1997, "A universal density profile from hierarchical clustering," ApJ, 490, 493

Oort, J. H. 1932, "The force exerted by the stellar system in the direction perpendicular to the galactic plane and some related problems," Bull. Astron. Inst. Neth., 6, 249

Paita, F., Celletti, A, and Pucacco, G. 2018, "Element history of the Laplace resonance: A dynamical approach," A&A, 617, A35

Plummer, H. C. 1911, "On the problem of distribution in globular star clusters," MNRAS, 71, 460

Poincaré, H. 1911, *Leçons sur les hypothèses cosmogoniques professées a la Sorbonne*, Paris: A. Hermann et fils.

Rabe, E., 1961, "Determination and survey of periodic trojan orbits in the restricted problem of three bodies," AJ, 66, 500

Rein, H. and Liu, S-H, 2012, "REBOUND: An open-source multi-purpose *N*-body code for collisional dynamics", A&A, 537, 128

Rickman, E. L., Ségransan, D., Marmier, M., *et al.* 2019, "The CORALIE survey for southern extrasolar planets. XVIII," A&A, 625, 71

Rosse, The Earl of, 1850, "Observations on the Nebulae," Phil. Trans. R. Soc., 140, 499

Sofue, Y., Tutui, Y., Honma, M., *et al.* 1999, "Central rotation curves of spiral galaxies," ApJ, 523, 136

Sollima, A., Martínez, D., Valls-Gabaud, D., *et al.* 2011, "Discovery of tidal tails around the distant globular cluster Palomar 14," ApJ, 726, 47

von Zeipel, H. 1910, "Sur l'application des séries de M. Lindsteadt à l'étude du mouvement des comètes périodiques," Astronomische Nachrichten, 183, 345

Wong, I., Shporer, A., Vissapragada, S., *et al.* 2022, "TESS revisits WASP-12: Updated orbital decay rate and constraints on atmospheric variability," AJ, 163, 175

Figure Credits

Astronomical images were obtained from a variety of sources with a preference for public-domain images or those licensed under Creative Commons. When we used figures from peer-reviewed scientific papers, we followed the publishers' guidelines for securing the necessary permissions.

Graphs composed by the technical editor were created using the Python matplotlib package in Jupyter notebooks. Artwork and schematics created by the author were composed using Microsoft PowerPoint.

Some plots employed orbit integrations using the galpy Python library for galaxy dynamics developed by Jo Bovy (Bovy 2015). Some of the solar system three-body orbit integrations were performed using the REBOUND Python/C package (Rein and Liu 2012).

Solar system data were obtained from the IAU Minor Planet Center, the NASA/JPL Horizons On-Line Ephemeris System (Giorgini, JD and JPL Solar System Dynamics Group, https://ssd.jpl.nasa.gov/horizons), and the Asteroids Dynamic Site (AstDyS-2, https://newton.spacedys.com/astdys).

Cover image: NASA *Hubble Space Telescope* image of the Antennae interacting galaxies (NGC 4038/39). Image size is 2.80 × 2.78 arcmin, corresponding to 15.5 × 15.4 kpc at the distance of the Antennae (d = 19 Mpc). Composite of visible and near-infrared images from *Hubble*'s Wide Field Camera 3 and the Advanced Camera for Surveys. [NASA, ESA, and the Hubble Heritage Team (STScI/AURA)-ESA/Hubble Collaboration]

Dedication: Presentation scene from *The Book of the Queen*, ca. AD 1410–1414, a collection of the works of Christine de Pizan commissioned by Queen Isabeau of France. Miniature attributed to the Master of the Cité des Dames and workshop. Harley MS 4431 fol. 178r, held and digitized by the British Library.

Figures 1.1–1.3, 2.1: Drawn by the author.

Figures 2.2, 2.3: Computed and plotted by the technical editor.

Figure 2.4: Plotted by the technical editor, using orbit elements for the Jupiter trojans from the IAU Minor Planet Center MPCORB database, and osculating elements for Mars and Jupiter computed using the JPL Horizons System.

Figures 2.5, 2.6: Reproduced from Rabe 1961, Figure 1. © AAS.

Figure 2.7: Plotted by the technical editor, using orbit positional data for 2013 BS$_{45}$ and Earth computed using the JPL Horizons System.

Figures 2.8, 2.9: Drawn by the author.

Figure 2.10: Plotted by the technical editor, using data calculated by Todd Thompson.

Figure 3.1: Plotted by the technical editor, using data from the IAU Minor Planet Center MPCORB database.

Figure 3.2: Drawn by the author.

Figure 3.3: Plotted by the technical editor, using osculating orbit elements for the hildas from the IAU Minor Planet Center MPCORB database, and osculating elements for Mars and Jupiter computed using the JPL Horizons System.

Figures 3.4, 3.6: Reproduced from Murray and Dermott 1999. Axis labels have been modified for readability, with permission of the original authors.

Figures 3.5, 3.7: Calculated using REBOUND (Rein and Liu 2012) and plotted by the technical editor, adopting initial conditions of Murray and Dermott 1999.

Figures 3.8, 3.9: Plotted by the technical editor, using data from the AstDyS-2 database.

Figure 3.10: Plotted by the technical editor, using data from the IAU Minor Planet Center: MPCORB for asteroids and AllCometEls.txt for comets. Jupiter orbit elements are from the NASA/JPL Solar System Dynamics database.

Figure 3.11: Drawn by the author, following Paita *et al.* 2018.

Figures 4.1–4.3, 5.1: Drawn by the author.

Figure 5.2: Plotted by the technical editor, using data from Birkby *et al.* 2017, downloaded from VizieR.

Figure 5.3: Plotted by the technical editor, using data from Moutou *et al.* 2009 and Rickman *et al.* 2019, downloaded from VizieR.

Figures 5.4, 5.5: Drawn by the author.

Figure 5.6: Plotted by the technical editor, using data from the NASA Exoplanet Archive.

Figure 5.7: Plotted by the technical editor, using data from Wong *et al.* 2022 and references therein, provided in digital format by Ian Wong.

Figure 5.8: Plotted by the technical editor, using data from (a) Dawson *et al.* 2014, provided in digital format by Rebekah Dawson, and (b) Holczer *et al.* 2016, dowloaded from VizieR.

Figure 6.1: Calculated and plotted by the technical editor.

Figure 6.2: Replotted by the technical editor from online versions of the data in Figures 1a and 3a in Sofue *et al.* 1999, downloaded from the public RC99 repository at the University of Tokyo.

Figures 6.3, 6.4: Calculated and plotted by the technical editor.

Figure 6.5: Drawn by the author.

Figures 6.6, 7.1, 7.2: Calculated and plotted by the technical editor.

Figure 7.3: Calculated using galpy and plotted by the technical editor.

Figure 7.4: Drawn by the author.

Figure 7.5: Reproduced from Figure 11, Plate II of Lissajous 1857.

Figure 7.6: Calculated using galpy and plotted by the technical editor.

Figure 8.1: Calculated and plotted by the technical editor.

Figure 8.2: Plotted by the technical editor, using data from de Boer *et al.* 2019.

Figure 8.3a: Image of M51 drawn by William Parsons, third Earl of Rosse [courtesy of Birr Castle Archives and the Trustees of Oxmantown Settlement Trust]

Figure 8.3(b): Image of M51 taken by the Advanced Camera for Surveys on the *Hubble Space Telescope*. NASA, ESA, S. Beckwith (STScI), and the Hubble Heritage Team (STScI/AURA). Converted to grayscale by the author and technical editor; image plotted with north to right.

Figures 8.4, 8.5: Calculated and plotted by the technical editor.

Figure 9.1: Image of the Antennae Galaxies taken by Robert Gendler (esahubble.org image heic0812c). Image cropped and converted to grayscale by the author and technical editor; image plotted with north to right.

Figure 9.2: Drawn by the author.

Figures 9.3, 9.4: Calculated and plotted by the technical editor.

Figure 9.5: Drawn by the author.

Figure 9.6: Image of the Cartwheel Galaxy and companion galaxies taken by NIRCam9 on the *James Webb Space Telescope*. NASA, ESA, CSA, STScI, and Webb ERO Production Team. Converted to grayscale by the author and technical editor.

Figure 9.7: Plotted by the technical editor, using data from Harris 1996, Baumgardt and Hilker 2018, and Baumgardt *et al.* 2019.

Figure 9.8: Reproduced from Figure 2 of Bonaca *et al.* 2020, by permission of the AAS.

Figure 10.1: Reproduced from Figures 10 and 11 of Martinazzi *et al.* 2014. Axis labels have been modified for readability, with permission of the original authors.

Figure 10.2: Plotted by the technical editor, using data from (a) de Boer *et al.* 2019 and (b) Baumgardt and Hilker 2018.

Figure 10.3: Reproduced from Figure 1 of Cohn 1980, by permission of the AAS.

Figure 10.4: Reproduced from Figure 3 of Hut and Bahcall 1983, by permission of the AAS.

Figures 10.5, 10.6: Calculated and plotted by the technical editor.

Further Reading, Bibliography, and Figure Credits header: Decorative element following the verse Matthew 7:8, from the *Book of Kells*, ca. AD 800. Scribe and artist unknown. Trinity College Dublin MS 58, fol. 48r. [© The Board of Trinity College Dublin]

Index